Chemical Interactions

Developed at
The Lawrence Hall of Science,
University of California, Berkeley
Published and distributed by
Delta Education,
a member of the School Specialty Family

© 2019 by The Regents of the University of California. All rights reserved. No part of this book may be reproduced or transmitted in any form or by any means, electronic or mechanical, including photocopying or recording, or by any information storage and retrieval system, without prior written permission.

1558519
978-1-62571-789-4
Printing 1 – 1/2018
Webcrafters, Madison, WI

Table of Contents

Readings

Investigation 2: Elements
Elements. 3
Substances on Earth. 13
Elements in the Universe. 15

Investigation 3: Particles
Particles 24
Three Phases of Matter. 28

Investigation 4: Kinetic Energy
Particles in Motion. 33
Expansion and Contraction. 40

Investigation 5: Energy Transfer
Energy on the Move. 46

Investigation 6: Thermos Engineering
Engineering a Better Design. 56

Investigation 7: Solutions
How Things Dissolve 64
Concentration 74

Investigation 8: Phase Change
Rock Solid. 89
Heat of Fusion .101

Investigation 9: Reaction
Better Living through Chemistry 110
How Do Atoms Rearrange? 118
Fireworks 130
Antoine-Laurent Lavoisier: The Father
    of Modern Chemistry. 134
Organic Compounds 141

Investigation 10: Limiting Factors
Careers in Chemistry. 148
Element Hunters 155

Images and Data 161

References
Science Practices. 182
Engineering Practices. 183
Engineering Design Process. 184
Science Safety Rules 185
Glossary 186
Index 190

The river is made of water, which is made of hydrogen and oxygen. The rock is made of minerals, which are often made of silicon, oxygen, and iron. The Sun, clouds, and trees are also made of fundamental substances called elements.

Elements

Have you ever wondered what things are made of?

For example, a table might be made of wood and nails. Your bread might be made of wheat, salt, and water. But what are wood and nails made of? What makes up wheat, salt, and water? What makes up you and everything around you?

About 2,000 years ago, people were asking the same questions. One idea was that everything was a mix of four basic properties: hot, cold, wet, and dry. They thought if you had just the right mix of hot and dry, that would make rock. A little less hot and a bit of wet might make wood. The right amount of all four properties might make a leaf.

Pure samples of the four properties were believed to be air, fire, earth, and water. These four things were thought to be the **elements** that make up everything else.

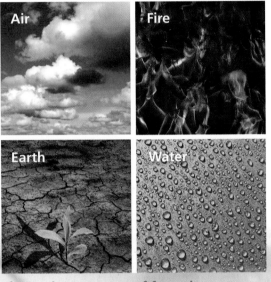

The ancient concept of four elements was the cornerstone of science, philosophy, and medicine for centuries.

By the 1800s, some people had a different idea about what things are made of. Chemists were busy investigating many **substances**. They heated substances as hot as they could. They put acid on them. They ran electric currents through them. Sometimes the substances separated into new substances. When this happened, scientists tested the new substances with heat, acid, and electricity. Some substances would not change any more. They called the unchangeable substances elements. The new elements had different names than the ancient elements, and there were many more of them. The new elements had names like iron, copper, carbon, oxygen, sulfur, and gold.

An element is a **fundamental** substance. It cannot be broken into simpler substances. Elements are the building blocks of **matter**. They combine to form all the different substances in the world.

By the middle of the 1800s, about 60 elements had been discovered. A lot was known about them. Scientists knew some of their **chemical properties**, such as what other elements they combine with. They knew some of their **physical properties**, such as the **mass** of standard samples of the elements. When scientists listed the elements, they put them in order by mass, starting with the lightest element they knew about, hydrogen.

The 17th and 18th centuries saw the beginnings of modern experimental chemistry. Great advances in the study of metals and gases led to the definition of an "element."

Investigation 2: Elements

The First Periodic Table

In 1869, a Russian chemist named Dmitry Ivanovich Mendeleyev (1834–1907) was writing a book about the elements. He made a set of element cards. Each card showed an element's name, **symbol**, and everything that was known about it. He put the cards in one long row from lightest to heaviest, hydrogen to uranium.

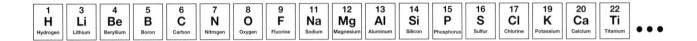

Mendeleyev saw something interesting in the line of element cards. The first two elements, hydrogen (H) and lithium (Li), had similar chemical properties.

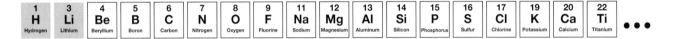

And as he looked down the line, he also noticed that sodium (Na) and potassium (K) had properties like hydrogen and lithium. The similar chemical properties showed up periodically, meaning at regular intervals, in his lineup.

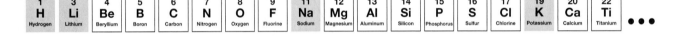

Then Mendeleyev saw that beryllium (Be), magnesium (Mg), and calcium (Ca) all had similar properties, but different from hydrogen and lithium. The similar chemical properties of beryllium, magnesium, and calcium showed up periodically, too.

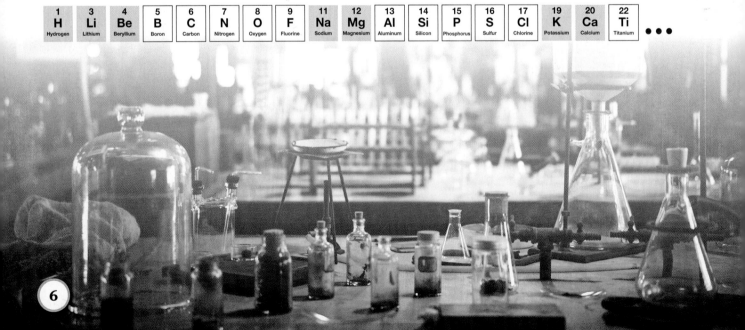

Mendeleyev had an idea. He reorganized the cards into several short rows. This way the elements with similar properties lined up in columns. The columns he created are now called groups. The periodic pattern of similar chemical properties gave the list its name, the **periodic table of the elements**.

When Mendeleyev had all the elements laid out, he noticed something was wrong. For instance, the chemical properties of titanium (Ti) were not like those of aluminum (Al) and boron (B) above it.

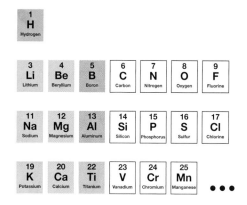

Dmitry Mendeleyev used cards to describe, arrange, and group the 63 known elements by the patterns of their chemical properties. The result was the first periodic table.

Titanium is a hard, shiny, strong metal, widely used in the aerospace industry. Shifting its position in the periodic table was a breakthrough for Mendeleyev.

Mendeleyev tried all kinds of things to make the cards line up better. He began moving elements around. When Mendeleyev moved titanium and its neighbors to the right, two things happened. The chemical properties of the elements lined up better. And there was a gap in the table of elements.

Mendeleyev **predicted** that an undiscovered element would fill that gap. He also predicted the properties of the new element. By lining up the known elements by their properties, Mendeleyev predicted about 30 new elements. Over the next 30 years, most of them were discovered.

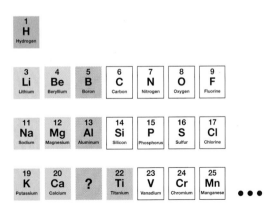

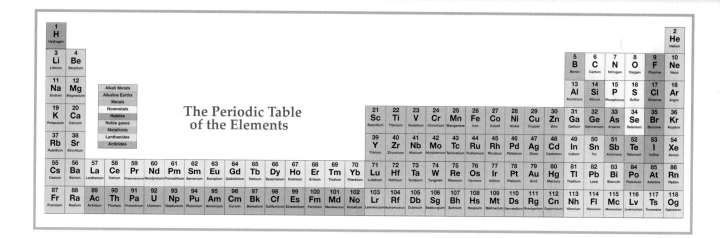

The Modern Periodic Table of the Elements

The modern periodic table of the elements organizes and displays all the elements, from simplest to most complex. Hydrogen, the simplest element, is number 1. Mendeleyev's idea of putting the elements in rows under each other, so that the chemical properties are similar in the columns, is still used. There are 2 elements in row 1, 8 elements in rows 2 and 3, 18 elements in rows 4 and 5, and 32 elements in rows 6 and 7. This is the modern periodic table of the elements.

This layout makes the table very long. Often 28 of the elements are pulled out and shown below the others. Then the table fits better on a standard piece of paper.

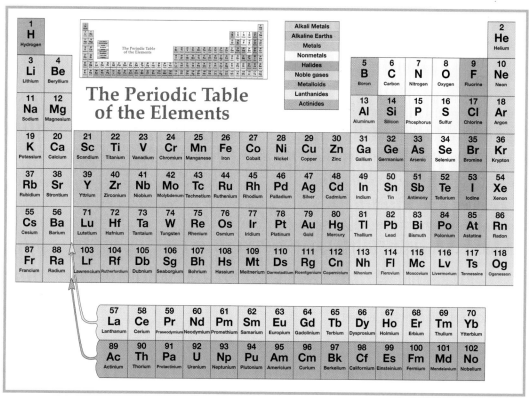

This is the modern periodic table of the elements. Elements in the same column have similar properties.

Investigation 2: *Elements*

Party balloons are often filled with helium, an element. Helium does not react with other elements and is less dense than air, so the balloons will float.

Color helps show which elements have similar chemical properties. In the periodic table used in this course, elements that are color-coded with orange, light blue, dark blue, light lavender, dark lavender, and yellow-green are **metals**. Green, red, yellow, and aqua elements (plus hydrogen) are nonmetals. The green elements on the right are called **noble gases**. They are interesting because they do not react with other elements.

Think Questions

1. What is an element?
2. How are matter and elements related?
3. How was Mendeleyev able to predict the existence of elements that had not yet been discovered?
4. What is the periodic table of the elements?

Sir Humphry Davy

Sir Humphry Davy (1778–1829) was born in Cornwall, England. As a young man, he studied medicine. But his life changed when he picked up a book about chemistry.

Davy began conducting experiments in his small laboratory. Some of his first efforts ended in explosions. Others filled his lab with strange **gases**. But Davy's knowledge of chemistry grew. He became a university teacher when he was 24 years old.

Davy became interested in separating substances until they could not be separated any more. He used a battery to run electricity through a **solution** of **potash**. Potash is potassium carbonate, K_2CO_3, a substance used to make glass and soap. When electricity ran through the potash solution, the potash separated.

That was how Davy discovered his first element, potassium (K). People say that he actually danced around the room after this discovery. Davy went on to become one of the greatest element finders of all time. He discovered more elements than anyone else! Using his electricity methods, Davy discovered seven elements: sodium (Na), magnesium (Mg), boron (B), potassium (K), calcium (Ca), barium (Ba), and chlorine (Cl).

Using the recently invented (1800) battery, Davy developed the process of electrolysis to isolate and identify new elements. The field of electrochemistry was born.

Marie Curie

Marie Sklodowska Curie (1867–1934) was born in Warsaw, Poland. In 1891, she moved to Paris, France. There she studied mathematics, physics, and chemistry, and completed her degree in physics.

She set up a small lab in the basement of the school where her husband, Pierre, taught. She studied the **radiation** coming from uranium ore. While analyzing the ore, she discovered two new elements, polonium (Po) and radium (Ra), in the ore sample. The samples of radium she produced constantly glowed green. She invented the term *radioactivity* to describe the radiation given off by the elements.

Curie, her husband, and Antoine-Henri Becquerel (1852–1908) were awarded the Nobel Prize in Physics in 1903 for their research on radiation. She was the first woman ever to win. She was awarded the Nobel Prize in Chemistry in 1911 for the discoveries of polonium and radium. She was the first person ever to win twice!

During World War I (1914–1918), Curie trained people to use X-rays to find bullets in wounded soldiers. Unfortunately, Curie did not know that radiation is dangerous. In 1934, she died from an illness caused by exposure to radioactive materials. Her notes and lab equipment are still radioactive today, more than 100 years after she did her research.

Marie Curie conducted groundbreaking research at a time when women were not usually allowed to pursue science. She is the only person ever to win a Nobel Prize in two categories.

Substances on Earth

How many elements on the periodic table do you recognize?

Many elements occur naturally. They can be found rather easily on Earth. For example, the lead in a drawing pencil is pure carbon. Carbon is element 6, also known by its symbol, C. Charcoal is another form of carbon and is also a good drawing tool.

Humans dig into rocks for natural resources such as gold (79, Au), silver (47, Ag), iron (26, Fe), and aluminum (13, Al). You've probably encountered these metals, perhaps in the form of a gold ring, silver earrings, a cast-iron skillet, or aluminum foil.

Carbon and metals are not the only commonly found pure elements. A glowing sign in a store window might hold neon gas. Neon (Ne) is element 10. In a hospital, you might have noticed a tank with tubes to provide oxygen to patients. Oxygen (O) is element 8.

Gold and silver, along with copper, tin, and platinum, are among the few commonly found metal elements.

Few elements are found in **liquid** form on Earth. One liquid element is mercury (80, Hg). It is the silver liquid used in old **thermometers** and barometers.

What about everything else around you? The paper this article is printed on, the white substances you tested in class, the toothpaste you used this morning—what are those made of? They are made of elements, like everything else in the universe. But they are not made of just one element, like the pencil lead, an iron skillet, or neon gas. They are a combination of several elements that form new substances.

You already identified the elements that make up the substances you used in class. They include sodium chloride (sodium and chlorine), calcium carbonate (calcium, carbon, and oxygen), and magnesium sulfate (magnesium, sulfur, and oxygen). Many, many substances are made of more than one element. For example, glass is made of silicon and oxygen, steel is made of iron and chromium, and brass is made of copper and zinc.

> **Think Question**
>
> 1. How do elements combine to form the varied substances in the classroom and throughout the universe? Write your ideas now, and keep thinking about this question as you go deeper into this course.

Most elements are found in combinations. These colored pencils are made of different combinations of elements to produce unique colors.

Just as 26 alphabet letters can form thousands of words, a mere 90 naturally occurring elements make up everything on Earth and in the vast universe.

Elements in the Universe

All across the universe, huge stars explode and send giant clouds of gas and dust into space.

About 5 billion years ago, in a region that would become the solar system, gravity pulled some of that gas and dust together. The small bits combined to become larger and larger over many millions of years, finally forming the Sun, the planets, and everything on them. Everything in the world, including you, is made of stardust.

What was that stardust made of? Elements. **Particles** of all 90 naturally occurring elements flew around in that space cloud. And when Earth formed, all 90 of those elements became part of Earth.

Investigation 2: Elements

Elements in the Sun

Our star, the Sun, has mass and occupies space. It is matter. All matter is made of elements. The Sun is no exception.

If you could separate the Sun's elements and organize them in a pie chart, the wedge of hydrogen would be the largest. The Sun is about 75 percent hydrogen by mass. The next largest wedge would be helium. The Sun is about 23 percent helium. The next three elements are very small wedges: 0.9 percent oxygen, 0.3 percent carbon, and 0.1 percent **nitrogen**. The Sun is mostly hydrogen and helium with small amounts of other elements.

Take Note

What elements do you think make up Earth? The ocean? The atmosphere? Your body? In your notebook, record your predictions.

 Access "Top 10" in the "Periodic Table of the Elements" online activity on FOSSweb.

Sun Elements

- Hydrogen 75%
- Helium 23%
- Oxygen 0.9%
- Carbon 0.3%
- Nitrogen 0.1%

This chart shows the most abundant elements in the Sun.

Earth Elements

The planet we live on, Earth, is made of all 90 naturally occurring elements. But the elements are not equally abundant. Some elements are common, while others are extremely rare. Those elements in the periodic table that you have never heard of are the rare ones.

Earth and the Sun have different amounts of the elements. The most abundant element on Earth is iron (35 percent). The massive **core** of the planet is mostly iron. Next are oxygen (28 percent), magnesium (17 percent), and silicon (13 percent). These three elements are the main elements in minerals and rocks. They also make up the largest part of the planet: two outer layers called the **mantle** and the **crust**.

Rocks and minerals make up all the layers of Earth: core, mantle, and crust. The color of a rock or mineral can be a clue to the elements it contains.

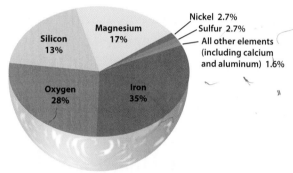

This chart shows the most abundant elements in Earth.

Investigation 2: *Elements*

The other major elements that make up Earth are nickel (2.7 percent), sulfur (2.7 percent), calcium (0.6 percent), and aluminum (0.4 percent). The remaining 82 elements together make up a tiny 0.6 percent of Earth.

It might seem that Earth is a pretty simple place. It is made mostly out of a dozen or so elements. But Earth's composition is not simple. It is not the number of elements that determines how complex things are. It is the ways the elements combine to make different substances. These few elements can combine to make millions of different materials. That's where the wonderful variety on Earth comes from.

Elements in the Ocean

The ocean covers almost three-quarters of Earth's surface. And the ocean is deep. That adds up to a lot of water. Water (H_2O) is made of two elements—hydrogen and oxygen. For this reason, the ocean is 85 percent oxygen and 11 percent hydrogen by weight.

Sea water is salty because it contains a lot of other elements. The most common salt is sodium chloride (NaCl). As a result, chlorine (2 percent) is the third most-abundant element in the sea, and sodium (1 percent) is the fourth most-abundant. Many other kinds of salt are **dissolved** in the sea, too.

All 86 of the other elements found on Earth are in the sea as well. For billions of years, water has been washing across the land, flowing into rivers, and making its way to the sea. All of the elements dissolved by water on land end up in Earth's ocean. That means there is actually some gold in sea water. But the amount of gold is so small that it wouldn't be worth the expense to get it out.

Both Earth's land and Earth's ocean contain the same elements, all 90 of them, though most are found in trace amounts.

Even the invisible air around us is matter made up of elements. Nitrogen is by far the most abundant element in Earth's atmosphere. Clouds are made of water droplets, so they consist of hydrogen and oxygen.

Elements in the Sky

The **atmosphere** surrounding Earth is matter in its gas **phase**. Matter is made of elements. What elements are in the atmosphere?

The atmosphere is mostly nitrogen (78 percent). The second most-abundant element is oxygen (21 percent). The third most-abundant element is the noble gas argon. Less than 1 percent of the atmosphere is argon.

The other elements in the air are present in small quantities. Water in the air has the element hydrogen. **Carbon dioxide** gas in the air has the element carbon. Smoke from fires and exhaust from industries and motor vehicles add to the air. Some exhaust adds carbon dioxide to the atmosphere, which affects Earth's climate. Other exhaust contains elements that can damage plants and animals. Mercury, lead, and some substances containing sulfur and nitrogen can be health hazards.

Elements in You

You are made of elements in the periodic table. How many of the 90 elements do you think it takes to make a human?

You probably have a trace amount of every element in your body. That is because elements are found everywhere, including our air, water, and food. For instance, helium is in the air in tiny amounts. Small amounts of helium enter our bodies when we breathe. We do not need helium to survive, but it appears in our bodies anyway.

Other elements are essential for life. We need tiny amounts of some elements, like chlorine and iodine, to stay healthy. But we need large amounts of others, such as carbon and oxygen.

Our natural interactions with the world around us—breathing, eating, drinking—produce an ongoing exchange of elements between our bodies and our surroundings.

Investigation 2: Elements

The human body is about 75 percent water. Much of the rest of the mass of your body is made from the element carbon combined with other elements. Carbon, hydrogen, and oxygen combine to form **carbohydrates** (sugars and starches), **lipids** (oils, fats, and waxes), and **proteins**. Proteins also contain nitrogen. Skin, muscle, fat, and organs are made of carbohydrates, lipids, and proteins.

The tough, rigid parts of the body, like teeth, bones, and cartilage, are rich in the element calcium. Blood contains a lot of iron. Potassium and sodium are needed for nerve and brain function.

Quite a few elements make up our bodies. But when you add it all up by mass, about 98.5 percent of the human body is composed of only six elements. They are oxygen, carbon, hydrogen, nitrogen, calcium, and phosphorus. The remaining 1.5 percent is small amounts of a lot of different elements.

Human Body Elements

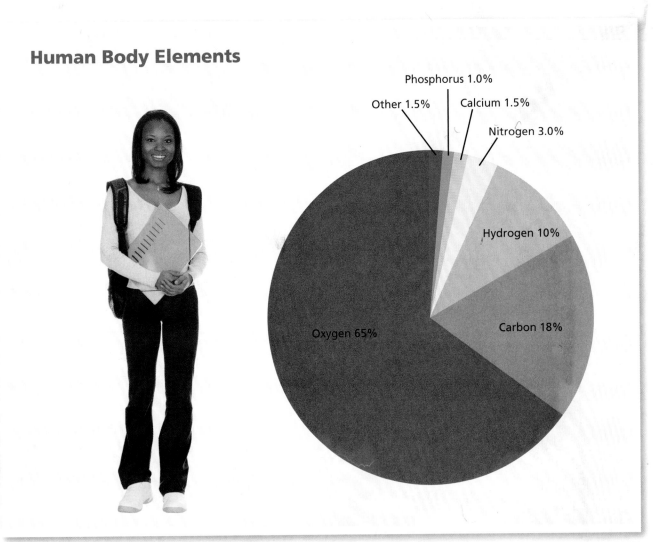

Even you are made of elements! Your trillions of cells are mostly water. About two-thirds of your body weight is oxygen, which combines with hydrogen to form water.

Trees and other plants are mostly carbon, hydrogen, and oxygen. So are honeybees, hummingbirds, horses, humans, and any other organism you can think of!

Most living organisms are chemically very similar to humans. So it is pretty safe to say that all life is based on the same six elements. Trees and other plants are mostly carbon, hydrogen, and oxygen, just like humans. So are the birds, insects, and spiders living in and among the plants. And anything we use that comes from organisms, such as wood, leather, paper, cotton, fabrics, plastics, food, and fuel, is also made of these elements.

So what elements are found in the universe? The answer is all of them. And which elements are abundant? Only a few. Common matter is made mostly of just a handful of elements.

Take Note

Go back to your predictions from the beginning of the article. Update the lists to include information from the article.

Think Questions

1. What element is among the five most abundant elements in the Sun, Earth, ocean, atmosphere, and organisms?
2. What does it mean when people say everything is made of stardust?
3. Why are the elements carbon, hydrogen, oxygen, and nitrogen important to life on Earth?
4. How can there be so many different substances in the world if only a few elements are common?

Investigation 2: *Elements*

The tiniest possible piece of an element that has all the properties of that element is a particle. Particles are too small to see. What do you imagine they would look like?

Particles

Everything on Earth is made of elements. The planet, the ocean, the atmosphere, and all life. But what are elements made of?

Elements are the building blocks of matter. On Earth there are 90 different elements. Carbon is one element. Aluminum is another element. Gold is a third element. All of the elements are different from one another. Each element has unique properties.

What Are Elements Made Of?

If you cut a copper wire in half, you have two smaller copper wires. If you cut each of those short wires in half, you have four really short copper wires (1). If you cut those in half, you have some little bits of copper that do not look like wire anymore. But they are still pieces of the element copper.

Imagine cutting one of those bits of copper into a million tiny pieces. Then cut one of those tiny pieces into a billion pieces (2). The pieces would be too small to see even with a microscope. But they are still pieces of the element copper.

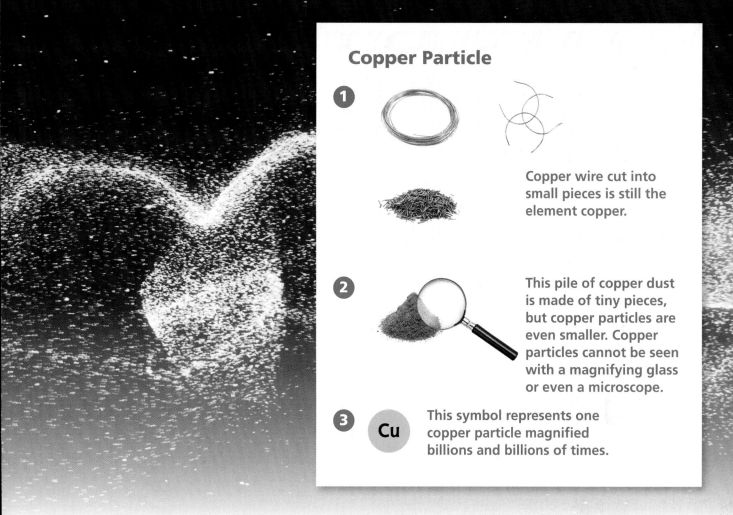

If you keep cutting the copper bits in half, you will finally end up with the smallest possible piece of copper (3). The smallest piece of copper is the copper particle. If you cut the copper particle in half, it is no longer copper. We can represent the copper particle as a ball labeled with the element's chemical symbol, Cu.

Unique Particles

We will use particle to describe the smallest piece of any substance that is still that substance. The element copper is made of copper particles. The element aluminum is made of aluminum particles. The element gold is made of gold particles. Each of the 90 elements is made of its own kind of particle.

Other substances you have worked with are not pure elements, but they are also made of particles. The particles of elements combine to make millions of different substances. Each new substance is made of its own unique particle. This means that there are millions of different kinds of particles in the world! How can we represent particles when elements combine?

Investigation 3: *Particles*

Let's use water as an example. Two hydrogen (H) particles and one oxygen (O) particle can combine to form the substance water (H_2O). The **chemical formula** for water shows that the water particle is made of two hydrogen particles and one oxygen particle. Diagram 1 shows one way to represent the water particle.

Sodium bicarbonate is another substance. Like all substances, it has a unique particle. Its chemical formula describes the sodium bicarbonate particle. The chemical formula for sodium bicarbonate is $NaHCO_3$. Diagram 2 shows the element particles that combine to make one sodium bicarbonate particle, which are one sodium (Na) particle, one hydrogen (H) particle, one carbon (C) particle, and three oxygen (O) particles.

The sodium chloride (NaCl) particle is small. It includes just two particles, one of sodium (Na) and one of chlorine (Cl). Magnesium sulfate ($MgSO_4$) is a larger particle, made of one magnesium (Mg) particle, one sulfur (S) particle, and four oxygen (O) particles. The sucrose ($C_{12}H_{22}O_{11}$) particle is the largest of the substances you investigated.

Take Note

How many particles of carbon (C) are in the sucrose particle? How many particles of hydrogen (H)? Oxygen (O)? What is the total number of element particles that combine to make up the sucrose particle?

Particles of Common Substances

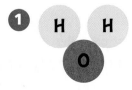

1. The elements hydrogen and oxygen combine to form a water particle.

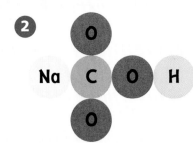

2. The elements sodium, carbon, oxygen, and hydrogen combine to form a sodium bicarbonate particle.

Representing Particles

Particles of substances can be lots of different shapes. Some particles are long and thin. Some are bumpy. But even the largest particles are far too small to see.

Even though particles are different sizes and shapes, you can model them as arrangements of little balls. Each ball represents one particle of an element, labeled with the element's chemical symbol. You can draw the little balls in different colors for each element to highlight the differences. The little balls can be connected to represent the particle of any substance you can think of.

Think Questions

1. What is a particle?
2. What is the difference between an element and a particle?
3. How many different kinds of particles are there in the world? Explain your answer.

Just as everything you eat is made of different combinations of ingredients, those ingredients are made of different combinations of element particles.

Three Phases of Matter

Penguins live among all three phases of matter: solid, liquid, and gas.

The **solid** ice is a secure place for these birds to stand. The liquid water is where they swim to find food. And the invisible gas (air) is blowing all around them.

Ice, water, and air have different properties. Solid ice has definite size (**volume**) and shape. In other words, the size and shape do not change. That is characteristic of all solid substances.

The penguins are surrounded by water in its three phases: solid ice, liquid water, and water vapor gas. Have you ever experienced all three at the same time and place?

Liquid water has definite volume, but its shape changes. It forms waves, crashes up on shore, and flows back to the sea. The volume of liquid water is always the same, but its shape depends on where it is. That is characteristic of all liquid substances.

Air does not have definite volume or shape. Air fills in around everything and can take any shape. The ability to change volume and shape is characteristic of all **gaseous** substances.

Why is ice solid? Why is water liquid, and air gas? The phase of a substance depends on the relationship between the particles of that substance. In solid ice, the particles are stuck together. They cannot move around because **forces** attract and hold the particles to each other. These attractive forces are called **bonds**. They "glue" particles together. When particles cannot move, the substance cannot change size or shape.

In liquid water, bonds hold particles close to each other, but they are not stuck together.

The particles move over and around each other. That allows water to flow. Particles in a liquid are packed close together, so the volume always stays the same. But because particles can move, the shape of a liquid can change. Liquids take the shape of their containers.

Properties of the Three Phases

Solids have definite volume and shape and sit on the bottom of their containers.

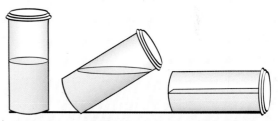

Liquids have definite volume, but their shape changes to fill the bottom of their containers.

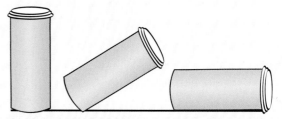

Gases have no definite volume or shape and always fill their containers.

In gas, particles fly around in space as individuals. Particles have a lot of space between them. Because they are not bonded, they spread out to fill a container. Gases have no definite volume or shape. Gases fill any container they are put into.

Applying Force to Matter

Let's look at three samples of matter. One sample is solid, one is liquid, and one is gas. Each sample is made of 39 particles. Each is in an identical container. This is how the particles would be organized in the three samples.

A syringe is a good tool for applying force to samples of matter. We can use a syringe to see what happens when force is applied to a solid, a liquid, and a gas.

This liquid has bubbles of gas floating through it. The liquid and the gas have different properties.

Three Samples of Matter

Solid

Liquid

Gas

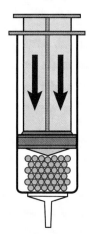

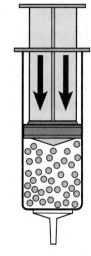

The particles of a solid are touching and bonded tightly.

The particles of a liquid are touching, but they can move.

The particles of a gas have a lot of space between them.

Solids and liquids are not very interesting in a closed syringe. They cannot be forced into a smaller space, because the particles are already in contact with one another. But there is a lot of space between gas particles. Gas can be **compressed**. The particles of compressed gas are forced closer together.

But there is a limit to how much gas can be compressed. At first, it is easy to push the plunger down. The air inside the syringe feels spongy. But the farther down you push the plunger, the harder the air feels. Why is that?

The air particles are flying around very fast inside the syringe. When they crash into the plunger and the walls of the syringe, they apply a force (1).

When the plunger pushes gas into a smaller space, it pushes the particles much closer together. This results in more particles hitting the plunger every second. The particles pushing on the plunger tip make it harder and harder to push the plunger (2).

 You can explore a model of this motion using the "Gas in a Syringe" online activity on FOSSweb.

Particles in a Plunger

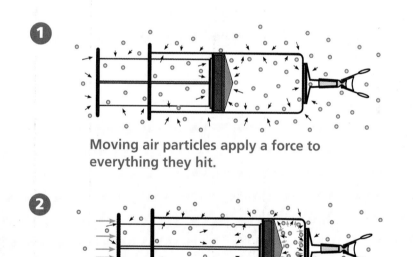

Moving air particles apply a force to everything they hit.

The force pushing the plunger down (blue arrows) pushes the particles of air in the syringe closer together.

A Bubble in a Syringe

A plastic bubble filled with air crumples up when you compress the air around it. Can you figure out why?

When air compresses, the air particles hit everything with more force. They hit the plunger, the walls of the syringe, and the plastic surrounding the air in the bubble. The plastic pushes on the air inside the bubble. The air inside the bubble compresses, just like the air outside the bubble. The number of air particles inside the bubble stays the same. But the space occupied by those particles (the volume) is smaller. The plastic bubble crumples because its air volume is smaller.

Think Questions

1. What crumples a plastic bubble in a syringe when you apply force to the plunger?
2. How is the motion of particles in a solid, a liquid, and a gas different?
3. Why does air feel hard when you push on the plunger of a closed syringe?

Compressing an Air Bubble

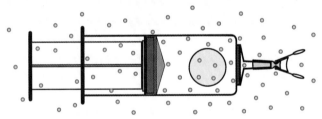

Before the plunger is pushed, the plastic bubble filled with air is spherical.

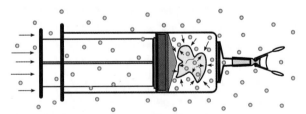

When the plunger is pushed, air is compressed in the syringe and the air bubble.

Air is a mixture of gases, mostly nitrogen and oxygen. The gas particles are always in motion, not just when the wind is blowing!

Particles in Motion

All substances are matter, made up of particles. Does this include invisible air?

Air is matter. It has mass and occupies space. Air is a **mixture** of many gases. It is made of approximately four-fifths nitrogen and one-fifth oxygen. All the other gases, including carbon dioxide (CO_2) and **water vapor** (H_2O), make up only a little more than 1 percent of the mass of a sample of air.

Air is matter in its gas phase. The nitrogen and oxygen particles in air are not bonded to other gas particles. Gas particles fly through space as individuals.

After you drink a bottle of water, you have an excellent container for an air investigation. The bottle looks empty, but it is full of air. Because air particles are flying all around, they are going in and out of the open bottle all the time. The **density** of air in the bottle is exactly the same as the density of the air outside the bottle. So every cubic centimeter of air inside and outside the bottle has the same number of particles.

Open Bottle

Air particles fly through space as individual particles. Air particles fill an open bottle.

Remember that we are using a model to understand particles. Air particles are millions of times smaller than they look in the illustrations. A cubic centimeter (cm³) of air actually has about one quintillion air particles! A quintillion is a one followed by 18 zeroes. (1,000,000,000,000,000,000). The illustrations are therefore not accurate. But they are a good model for thinking about air particles.

Particles Have Kinetic Energy

Air particles are incredibly small and are always moving. And they move fast. At **room temperature**, they move at about 300 meters (m) per second.

All moving objects have energy. This kind of energy is called **kinetic energy**.

Anything that is in motion has kinetic energy. That includes you bicycling to class, water falling down a waterfall, and an oxygen particle in the air.

Kinetic energy, like all forms of energy, can do work. Air particles do work when they crash into things. Air particles push on each other, on you, on the walls of containers, and on everything else around them. Every air particle crashes into another particle about 10 billion times every second!

The amount of kinetic energy an object has depends on its mass and speed. You cannot change the mass of an air particle, but you can change its speed. You can make air particles move faster by **heating** a sample of air. Heating particles increases their kinetic energy by increasing their speed.

Bicycle riders in motion have kinetic energy, and the faster they move, the more energy they have. It's the same with air particles. The faster the speed, the greater the energy.

Investigation 4: Kinetic Energy

Let's go back to the air investigation. Stretch a balloon over the top of the bottle full of air. Now the air is trapped inside the bottle-and-balloon system. No particles can get in or out. The density of air particles is the same in the bottle, in the balloon, and in the air surrounding the system, as shown in diagram 1.

Now place the bottle-and-balloon system in a cup of hot water. The hot water warms the air inside the bottle. Particles in the warm air start to move faster. After a few minutes, the bottle-and-balloon system looks like diagram 2.

Why did the balloon inflate? The hot water heated the air in the bottle. As a result, the air particles began moving faster. Faster-moving particles have more kinetic energy. Faster-moving particles hit each other harder, which pushes them farther apart. The illustration shows that particles of warm air inside the system are farther apart.

Bottle-and-Balloon System

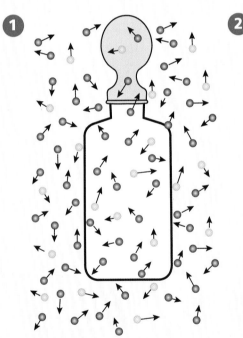

A balloon can trap the air inside a bottle.

Hot water increases the kinetic energy of the air particles inside the bottle-and-balloon system. The particles fly faster and hit each other harder. The particles push farther apart, causing the gas to expand.

Heating a Gas

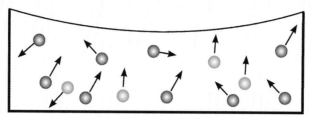

The particles in gases fly through space in all directions as individuals.

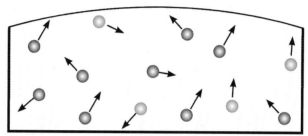

When gases get hot, the particles fly faster. Faster particles hit other particles harder, pushing the particles farther apart. This causes the gas to expand.

The faster-moving particles also push harder on the balloon. The force that the particles apply to the balloon is enough to stretch it. The increased kinetic energy of the particles pushes them farther apart (the air expands), and the balloon stretches to hold the increased volume of air.

Note that the particles themselves do not become larger or smaller! The relationship between particles changes, making the sample expand or contract, but the particles stay the same size.

When Gases, Liquids, and Solids Heat Up

Gas. Remember that the particles in a gas are not bonded (attached) to other particles. Each particle moves freely through space. When a sample of air heats up, the particles move faster and hit each other harder. The result is that the particles push each other farther apart. So when a container of gas has a balloon across the top, the balloon expands when the gas gets warm, because the heated gas requires more space.

Heating a Liquid

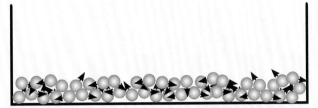

The particles in liquids are held close to each other. Particles bump and slide around and past each other.

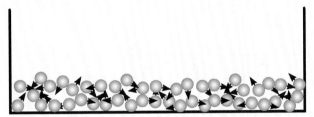

When liquids get hot, the particles bump and push each other more. This causes the liquid to expand.

Heating a Solid

The particles in solids are bonded. Particles move by vibrating, but they do not change positions.

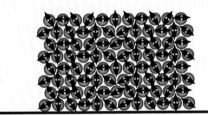

When solids get hot, the particles vibrate more. Increased vibration pushes the particles farther apart, causing the solid to expand.

Liquid. Particles in liquids are in close contact with one another. Attractions between the particles keep them from flying freely through space. The particles in liquids can still move over, around, and past one another. Individual particles in liquids can move all through the sample of liquid.

The motion of particles in a liquid is kinetic energy. When a liquid gets warm, the particles move faster. The particles have more kinetic energy. As a result, they hit other particles more often and hit harder. This pushes the particles farther apart. When particles are pushed farther apart, the liquid expands.

Solid. Particles in solids have bonds holding them tightly together. The particles cannot move around each other. But they can move a little. Particles in solids are always **vibrating** (moving rapidly back and forth) in place.

The vibrating motion of particles in solids is kinetic energy. Heating particles in a solid makes them vibrate faster, giving them more kinetic energy. Faster-vibrating particles bump into one another more often and hit each other harder. This pushes the particles farther apart, and the solid expands.

Heating and Cooling

When a sample of solid, liquid, or gas matter heats up, its particles gain kinetic energy. The increased kinetic energy pushes the particles farther apart. This causes the matter to expand.

When a sample of solid, liquid, or gas matter cools down, its particles lose kinetic energy. The decreased kinetic energy lets the particles come closer together. This causes the matter to contract.

Think Questions

1. What is kinetic energy?
2. How can you increase an object's kinetic energy?
3. What happens to a sample of matter when its particles lose kinetic energy?
4. How are particles in solids, liquids, and gases the same? How are they different?

Icicles are made of water in its solid phase. What does that tell you about its particles?

Expansion and Contraction

We are surrounded by solids, liquids, and gases. And they all warm up and cool down from time to time.

When they warm up, they expand. When they cool down, they contract. Sometimes **expansion** and **contraction** are useful. Other times they are a nuisance.

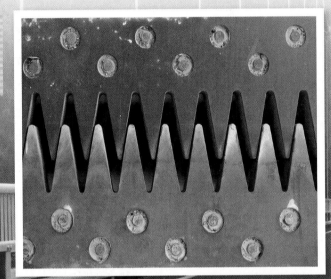

On a cold day, the bridge surface contracts. The fingers of the expansion joint don't overlap very much.

Bridges

This bridge is made of solid steel with a surface of solid concrete. During the summer, the surface of the bridge heats up. The surface expands. When the surface gets bigger, what will happen to the bridge? If the design does not allow the bridge to expand, it could buckle and break. Or the force of expansion could damage the steel structure.

Bridge **engineers** design a bridge's surface in sections. They place special joints where the sections meet. The bridge shown here has expansion joints with steel fingers that can move between each other. When the surface gets hot, the sections expand and get longer, and the fingers push between each other. When it gets cold, the sections contract and get shorter, and the fingers pull apart. The finger design allows cars to smoothly cross the junctions between the surface sections. Without these joints, bridges could become damaged as the **temperature** changes throughout the year.

Bridge expansion joints let the concrete roadbed adjust naturally to temperature changes without cracking or bending. Have you noticed these when driving over a bridge?

Thermometers

Thermometers are filled with liquid. The most commonly used liquid is alcohol. When the alcohol gets hot, it expands. When the alcohol gets cold, it contracts. Expansion and contraction of alcohol are useful properties for making a thermometer. Here's how it works.

If you want to know the temperature of the water in a vial, you place the bulb of a thermometer in the water. If the water is cooler than the bulb, energy transfers from the warm thermometer bulb to the cold water. As the alcohol cools down, the kinetic energy of the alcohol particles decreases. When the alcohol particles move less, they get closer together, and the alcohol contracts. When the alcohol contracts, it takes up less space inside the thermometer. The level in the thermometer stem goes down.

When you move the thermometer to a vial of warm water, the alcohol in the bulb heats up. The kinetic energy of the alcohol particles increases, and the volume of alcohol expands. The larger volume fills more of the thermometer stem.

By using the numbers marked on the thermometer stem, you can compare temperatures. These thermometers have marks that measure temperature in degrees Celsius (°C). The cool water in the picture is about 16°C. The warm water is about 51°C.

> **Take Note**
>
> How was the bottle system you built in class a model of an alcohol thermometer? What were some limitations of the bottle-system model?

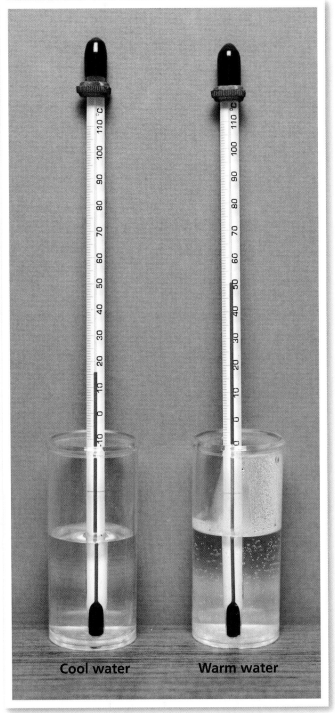

A thermometer contains alcohol, which expands when it gets warm and contracts when it gets cool. The height of the alcohol column indicates the temperature of the water around the thermometer bulb.

Microwave Ovens

Did you ever put a container in the microwave and forget to loosen the lid? You might have a food explosion! Here is the chain of events that can make that happen.

You want a quick breakfast, so you take some leftover scrambled eggs out of the refrigerator. You place the container in the microwave and hear a pop about 30 seconds later. You open the microwave to find a mess. The lid popped off, and took some of the egg with it. Why?

The air inside the container is gas. As the egg heats up in the microwave, so does the air in the container with the egg. The kinetic energy of the air particles increases, and they push harder on each other and on the lid of the container. If you look closely, you can see the lid bulging as the air warms. Eventually, the pressure created by the increased kinetic energy of the air particles will pop the lid right off the container, leaving you with a mess to clean up.

Expansion and contraction occur around us all the time. Air expands and contracts as it warms and cools. We do not see it happening, but we see the results. Weather, particularly wind, happens as a result of expanding and contracting air.

Eggs in a sealed container explode in a microwave because rapid heating causes particles of egg and air to move faster. When the air particles are moving fast enough, they can push off the container lid!

Investigation 4: *Kinetic Energy*

Structure Design

Have you ever been in an old, quiet building when there was a sudden creaking or cracking sound? Is it a haunted house? Probably not. Buildings expand and contract as they heat and cool, just like steel, thermometer alcohol, and air. When building materials expand or contract, beams, walls, and roofs move a little bit. When parts of a building move, they can make sounds.

Listen closely next time you are in a quiet building just after sunrise or just after sunset. Those are the times when you are most likely to hear unexpected sounds. Prepare an explanation to calm your scared little brother or sister!

Temperature change around a house can cause expansion or contraction of the building materials. Have you noticed this in your home?

Expansion joints in train tracks are intentional gaps that allow the steel tracks to expand as the temperature increases.

Train tracks are steel. Tracks expand when they get hot in the sunshine. Expansion joints keep them from buckling and bending. Next time you are close to some train tracks, look for these joints. What keeps sidewalks and roads from breaking and buckling? Their expansion joints are spaces between the concrete sections. The spaces are filled with material that can compress when the concrete expands.

Keep your eyes open when you are looking at very large structures and huge expanses of solid surface. To design these spaces, engineers had to consider the problems presented by expansion and contraction during heating and **cooling**. If you look closely, you can see how these structures were designed.

Think Questions

1. What are expansion joints, and why are they used?
2. What causes the lid to pop off a container of leftovers in the microwave?
3. Why does the alcohol level in a thermometer change at different temperatures?

Investigation 4: Kinetic Energy **45**

Energy on the Move

Have you ever burned your tongue on a cup of hot cocoa? Maybe you cooled it down by adding a splash of cold milk.

But why exactly does this work? When you mix cold milk and hot cocoa, what happens to the cold milk? And what happens to the hot cocoa?

After mixing, the cup of cocoa is warm throughout. It is cooler than the hot cocoa and hotter than the cold milk. It seems as if the hot cocoa gets colder and the cold milk gets hotter. The new temperature is between the starting temperatures of the cocoa and the milk. How does that happen?

One way to cool down hot cocoa is to add cold milk. Blowing on the hot surface can also speed up the cooling process.

Kinetic Energy

Objects in motion have kinetic energy. The particles of substances are always moving. So the particles that make up the hot cocoa and the cold milk have kinetic energy.

The amount of energy that a particle has depends on how fast it is moving. Faster-moving particles have more energy. Slower-moving particles have less energy.

The simple rule is that the more kinetic energy the particles have (the faster they are moving), the hotter the substance is. The particles in hot cocoa have more kinetic energy than the particles in cold milk.

This is important. *Kinetic energy of particles is directly related to temperature of a substance.*

> **Take Note**
>
> **Can you use your hands to create a model that shows the difference between a hot liquid and a cold liquid? Record a diagram of this in your notebook.**

Changing Kinetic Energy

Energy is **conserved**. The law of **conservation of energy** means that energy is never destroyed or created during interactions. The amount of energy in a system is always the same, but it can move from place to place. When energy moves from one place to another, it is called **energy transfer**.

Energy transfer happens when particles collide. Say a fast-moving particle hits a slow-moving particle. The slow-moving particle speeds up, while the fast-moving particle slows down. When a particle speeds up, it has more kinetic energy. When a particle slows down, it has less kinetic energy.

Energy transfers from a fast-moving particle to a slow-moving particle at the moment of impact. Look at the diagram.
1. Fast-moving particle A is on a collision course with slow-moving particle B.
2. The particles collide. At the moment of impact, energy transfers from particle A to particle B. As a result, particle A has less kinetic energy, and particle B has more kinetic energy.
3. The two particles are now moving at about the same speed. Energy transferred from particle A to particle B at the moment of impact.

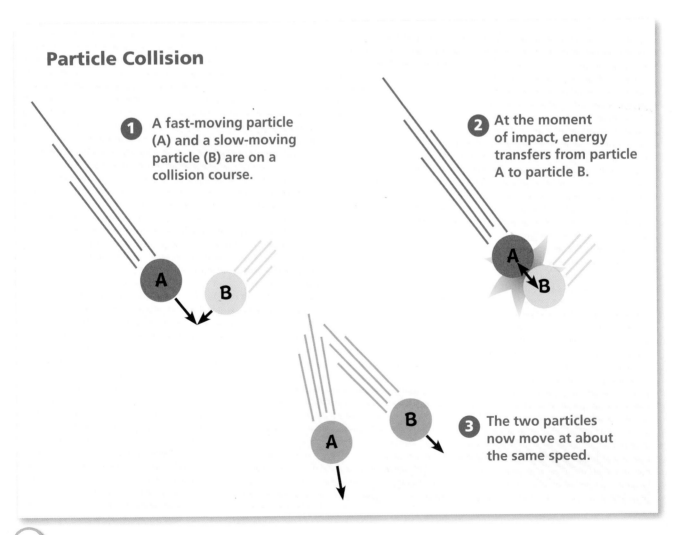

Particle Collision

1. A fast-moving particle (A) and a slow-moving particle (B) are on a collision course.
2. At the moment of impact, energy transfers from particle A to particle B.
3. The two particles now move at about the same speed.

The collision of two billiard balls is an example of energy transfer by contact.

The total kinetic energy of the two particles before the collision is exactly the same as the total kinetic energy after the collision. No kinetic energy is created or lost as a result of the collision, only transferred. The energy of the two-particle system is conserved.

But something did change. As a result of the collision, particle A has less kinetic energy, and particle B has more kinetic energy. The collision resulted in energy transfer from particle A to particle B. This kind of energy transfer, from particle to particle by contact, is called **conduction**.

Energy Transfer in Cocoa

The kinetic energy of the particles in hot cocoa is high. They are moving fast. The kinetic energy of the particles in cold milk is low. They are moving slowly.

When you pour cold milk into hot cocoa, the milk and cocoa particles start to collide. When a high-energy cocoa particle hits a lower-energy milk particle, energy transfers by conduction. The cocoa particles slow down, and the cup of cocoa cools down.

Look at the illustrations of the particle collision. Which particle represents a cocoa particle and which represents a milk particle? Can you see how the energy transfer reduced the speed of the cocoa particle? Remember, reduced particle speed means less kinetic energy. Lower kinetic energy means lower temperature.

 You can explore a model of this energy transfer using the "Mixing Hot and Cold Water" online activity on **FOSSweb**.

Investigation 5: *Energy Transfer*

Using a Thermometer

You can use a thermometer to find out if your cocoa is too hot. When you dip a thermometer in that cup of hot cocoa, it reads 90 degrees Celsius (°C). That is too hot. You need to add some cold milk. Put the thermometer in the cold milk. It reads 10°C. That should do the job. But how does the thermometer "know" that the cocoa is 90°C and the milk is 10°C? The answer is kinetic energy. The thermometer reports the average kinetic energy of the particles in a substance. That's what temperature is: average kinetic energy of particles.

How exactly does the thermometer work? Now that we have an idea of how energy transfer takes place, let's put the thermometer into the 90°C cocoa. The cocoa particles collide with the glass particles on the outside of the thermometer stem (1).

The glass particles gain kinetic energy and start vibrating more rapidly. The glass particles transfer energy to their neighbors by conduction, and those transfer energy to their neighbors. Pretty soon the whole glass stem is at 90°C (2).

The rapidly vibrating glass particles are in contact with the alcohol inside the thermometer. They transfer kinetic energy to the alcohol particles. Kinetic energy is conducted from alcohol particle to alcohol particle (3).

 Check out the "Thermometer" online activity on FOSSweb.

Thermometer Particles

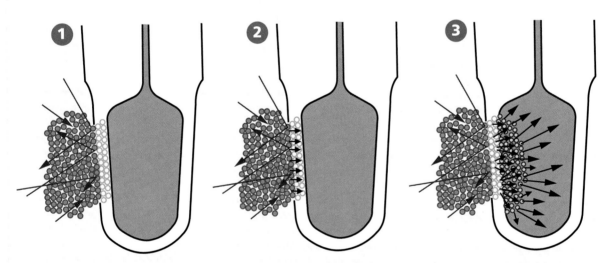

Energy transfers from cocoa particles to glass particles.

Energy transfers through the glass by conduction.

Energy transfers from the glass particles to the alcohol particles. The alcohol expands.

Whether measuring the temperature of cold milk or hot cocoa, a thermometer works the same way. Energy transfers from warmer to cooler materials until the alcohol level reaches a steady reading.

Soon all the alcohol particles are moving faster, pushing on each other more often and with greater force. The distance between particles increases, and the alcohol expands. As the alcohol expands, its volume increases. Alcohol pushes up the stem of the thermometer. The greater the kinetic energy of the alcohol particles, the more the alcohol expands. Energy transfers to the alcohol until the average kinetic energy of the alcohol particles is the same as the average kinetic energy of the cocoa particles. When this happens, the top of the alcohol is at the 90°C mark.

Now put the thermometer into the 10°C milk. Energy transfers from the outside of the glass thermometer stem to the milk. Energy transfers from glass particles, lowering the kinetic energy of all the glass particles. When alcohol particles collide with lower-energy glass particles, energy transfers from the alcohol to the glass. The alcohol loses kinetic energy and contracts. When the average kinetic energy of the alcohol particles is the same as the average kinetic energy of the milk particles, the alcohol level is at the 10°C mark.

Investigation 5: Energy Transfer

Energy Flows from High to Low

When two particles collide, is it possible for the faster-moving particle to end up going even faster? Can energy transfer from a low-energy particle to a higher-energy particle?

No. It never happens. Energy *always* transfers from a fast-moving particle to a slower-moving particle. As a result of an energy-transfer collision, the particle that was going faster before the collision will always be going slower after the collision. And the particle that was going slower before the collision will be going faster after the collision. Always.

It is sometimes useful to think of energy as flowing. Energy always flows from higher to lower, from hot to less hot (cold).

Equilibrium

When you pour cold milk into the hot cocoa, the higher-energy cocoa particles and lower-energy milk particles mix with one another. They collide with each other billions of times every second. The energy flows

What energy transfers occur when you eat hot pizza?

from the higher-energy cocoa particles to the lower-energy milk particles.

Almost instantly, the average kinetic energy of the milk particles is the same as the average kinetic energy of the cocoa particles. The kinetic-energy level is uniform throughout. It is lower than the cocoa and higher than the milk.

Has energy transfer between particles stopped? Not really. Even when the average kinetic energy of the mixture stays steady, some individual particles still have high energy, and some particles have low energy. But the number of high-energy particles is the same as the number of low-energy particles.

When the temperature is constant, the system is in a condition called **equilibrium**. At equilibrium, temperature does not change. When a mixture of hot cocoa and cold milk has reached equilibrium, you can use a thermometer to measure the equilibrium temperature. The equilibrium temperature is a measure of the average kinetic energy of all the particles in the system. This includes the cup, the mixture of cocoa and milk, and the thermometer.

Geothermal power plants create electricity from energy from Earth's core that has been transferred to the surface.

So has energy stopped flowing? Think about this. The phone rings and you talk to a friend for 20 minutes. Now the cup of cocoa is cold. Why? The room is cooler than the cocoa. Particles of air collide with the cup and the surface of the cocoa. Energy transfers from the cup of cocoa to the air. Energy continues to transfer to the air until the average kinetic energy of the cocoa is the same as the average kinetic energy of the air. We say the cocoa is room temperature. The cocoa is at equilibrium with everything else in the room.

Summary

All matter is made of tiny particles that are too small to see. The particles are in constant motion.

Objects in motion have kinetic energy. Particles are objects in motion, so they have kinetic energy. The faster a particle moves, the more kinetic energy it has.

Kinetic energy is related to temperature. The faster the particles in a substance move, the hotter it is. Temperature is a measure of the average kinetic energy of the particles in a sample of matter.

Energy can transfer from one particle to another when particles collide. Energy always transfers from a higher-energy particle to a lower-energy particle. The transfer of kinetic energy during contact is called conduction. Matter heats up and cools down because of energy transfer between particles.

Think Questions

1. Why do you think an ice cube feels cold when you hold it?
2. What will happen to the balloon stretched over the mouth of this "empty" bottle when the bottle is placed in hot water? Explain all the energy transfers.
3. When does energy transfer from a cold object to a hot object?

Investigation 5: Energy Transfer

Engineering a Better Design

You just sat down with a cup of hot cocoa when your phone rings. While you talk, energy transfers from the hot cocoa to the mug and air.

In 20 minutes, the average kinetic energy of the cocoa and the mug is the same as the average kinetic energy of the air. The cocoa is now at room temperature.

If you had put your hot cocoa in a thermos, it would still have been hot by the end of your conversation. How does the thermos keep hot liquids hot and cold liquids cold?

We have already learned that energy transfers from one particle to another when particles collide. This transfer of kinetic energy is called conduction. Energy always transfers from a high-energy particle to a lower-energy particle.

Depending on the length of the phone call, this hot cocoa may be room temperature cocoa before the next sip.

Insulation

When you use a thermos, the rate of conduction from the hot cocoa to the cool air decreases. Special materials in the thermos reduce energy conduction. These materials are called **insulation**.

Scientists learn about properties of insulation by developing theories and conducting tests. Their evidence and reasoning lead to models that explain why certain materials are good insulators. Scientists and engineers often work together to find creative solutions to an **engineering problem**. Engineers use scientific knowledge to design a solution to a real-life problem. In class, you used engineering design to determine which material is the best insulator.

First, you had to understand the **criteria** and **constraints** of the problem. Then, you came up with ideas and developed a testing plan. In each test, you measured how fast energy transferred from a liquid to the environment. The better the insulator, the slower the energy transfer. You used the results of your first test to improve your design. You tested that design, too. Engineers evaluate each possible solution. How well does it meet the criteria and constraints of the problem? They won't stop until they have a working solution.

Design of a Thermos

Engineers used a similar system to design a thermos. A modern thermos has a shiny liner. The outside layer is usually stainless steel or plastic. What you can't see is the space between the two layers. What special material insulates that space? Nothing! When the thermos is made, the air between the two layers is removed. Then the layers are sealed so nothing can get back in. This sealed empty space without any particles of air is a **vacuum**. Why is a vacuum the best insulator?

> **Take Note**
>
> Think about conduction, which is energy transfer between particles. Write down your idea about why a vacuum affects the rate of conduction.

Thermos Wall Particles

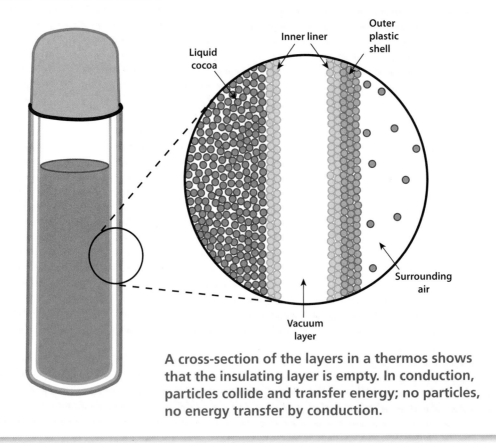

A cross-section of the layers in a thermos shows that the insulating layer is empty. In conduction, particles collide and transfer energy; no particles, no energy transfer by conduction.

A well-made thermos keeps hot tea hot all day. It works the same way—slowing energy transfer by conduction—to keep iced tea cold all day.

A vacuum has no particles to transfer energy between the liquid in the container and its environment. Energy can still transfer by conduction through the thermos lid and the walls of the container. But the vacuum layer slows the overall rate of conduction.

Eventually, the contents of any thermos will become the same temperature as the environment. Its insulation cannot completely stop energy transfer. But a good thermos can keep contents from becoming room temperature for hours.

Insulation in Buildings

We use insulation every day in many ways. For example, our winter clothes reduce energy transfer from our bodies to the cold environment.

Insulation is also important in buildings. In the average home, half of all the energy used is for heating and cooling the building. Insulation in walls, windows, and ceilings reduces energy transfer. A well-insulated home stays warm during cold months and cool during hot months. So well-insulated buildings are more energy efficient than poorly insulated buildings. People in these buildings do not need to use the heater or air conditioner as much, and that lowers energy use.

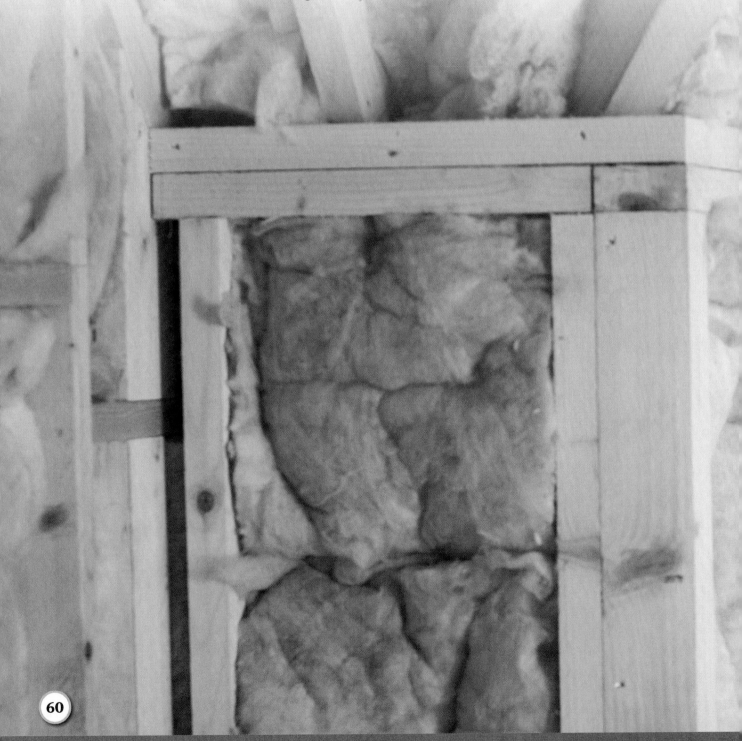

Engineers have to consider many factors when designing insulation for an entire building. Criteria include how large the building will be, the climate, and fire safety. Constraints include the cost of materials and work, health concerns related to materials, and construction deadlines.

Adding insulation to a home can make it more comfortable in the winter and summer, as well as lowering energy use.

Engineers can choose from many good insulating materials, as you learned in class. Some materials used for building insulation are brand new. Others are recycled, like sugarcane fiber or denim from old jeans.

Take Note

List some materials that you use in cold weather to reduce energy transfer from your body. Which of those materials did you or your classmates test in a thermos design?

Investigation 6: Thermos Engineering

Double-Paned Windows

The overall design of a building can also improve insulation. Making the walls thick is one simple technique. Another is using windows with double-paned glass. Double-paned windows insulate about twice as well as single-paned windows. Why is that?

Compare the mug and thermos again. Like the mug, a single-paned window rapidly conducts kinetic energy from a warm place to the cold surroundings as particles collide. Double-paned windows have a space between the two layers of glass, like a thermos.

The space between panes is not a vacuum, however. Window designs have criteria that make a vacuum unsuitable. Windows need to be strong and watertight. Engineers decided to fill the space between panes with a substance that has very few particles spread out over a lot of space. What substance meets those criteria? (Hint: Think about solids, liquids, and gases.)

When it is cold outside, it's nice to be cozy and warm inside. Double-paned windows help keep the interior of your home warmer by reducing energy transfer with outside particles.

The space in most double-paned windows is filled with a gas that is very stable, like argon (Ar). Some windows are filled with krypton (Kr). Using a gas between two panes reduces conduction and provides better insulation than a single, solid pane.

So engineers have developed a good solution, the double-paned window. Does that mean they will stop coming up with new ideas? No! Engineers are always improving designs. They are designing triple-paned windows for even better insulation. Your class designed insulation in just a few days. Imagine what you might develop if you kept testing new designs!

Think Questions

1. What criteria and constraints are important considerations when designing a thermos?
2. Fluffy materials are good insulators. What other criteria might affect the design of a winter jacket?

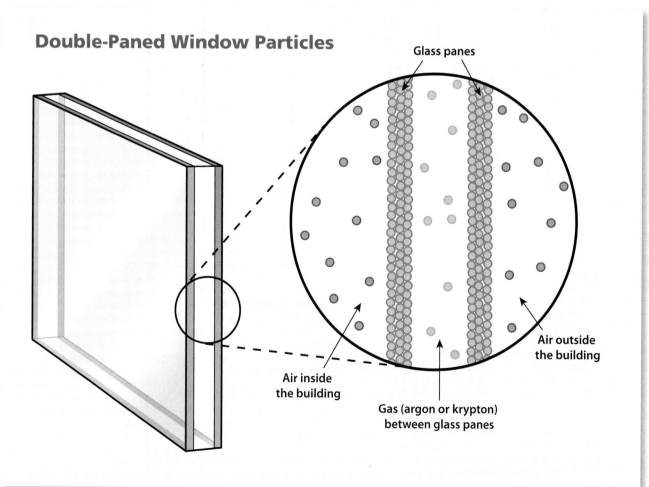

Double-Paned Window Particles

A cross-section of a double-paned window shows the insulating layer of gas sandwiched between two layers of glass.

Investigation 6: *Thermos Engineering*

How Things Dissolve

The sore throat and cough of a cold are very unpleasant. To get relief, you can pop a cough drop in your mouth.

The cough drop makes your throat feel a little better and puts the cough on hold for a while. But what happens to the cough drop? After a few minutes, it is a lot smaller than it was when you first put it in your mouth. And not long after that, it is gone. Where did it go?

The cough drop, which is mostly sugar, dissolved. It dissolved in saliva, which is mostly water. As it dissolved, the medicine flowed down your throat, bit by bit, soothing the pain.

Cough Drop in Water

Start at time 0   10 min.   20 min.   4 days

Given enough time, the red solid cough drop entirely dissolves in the clear liquid water. The particles of the cough drop disperse evenly and give the water a pink tint.

A cough drop contains ingredients to relieve cold symptoms. The cough drop is in solid phase, but will dissolve in your mouth.

Dissolving in Water

You can observe the dissolving process more easily by putting a red cough drop in a cup of plain water. Within a minute or two, a pool of red color forms around the cough drop. The red color is coming from the solid cough drop. Is the cough drop **melting**?

No, it is not melting. Melting is caused when energy transfers during heating. The cough drop in the cup of water is not being heated.

The cough drop is dissolving. It is breaking apart bit by bit, and the bits are moving into the water. In 10 minutes, the cough drop will be very small, and there will be a thin layer of red water on the bottom of the cup. In 20 minutes, the solid cough drop will be gone, and the red layer at the bottom of the cup will be larger. A day later, the red color will have moved higher in the cup, and in 4 days, the whole cup will be pink. The cough drop will be spread evenly throughout the water in the cup.

A Close Look at Dissolving

Water is made of water particles. They are in constant motion, bumping and banging around and over one another. But the water particles don't just bump into one another. They bump into everything that is in the water. If you put a cough drop in a cup of water, water particles will bang into it billions of times every second (1).

The cough drop is a solid, so its particles (shown in red) are held together by attractive forces called bonds. The bonds keep the cough drop from falling apart in the package before you use it. Then you drop the cough drop into water, a liquid. Water particles (shown in blue) hit the sugar particles on the edge of the cough drop and transfer enough energy to break those bonds (2).

Cough Drop Dissolving

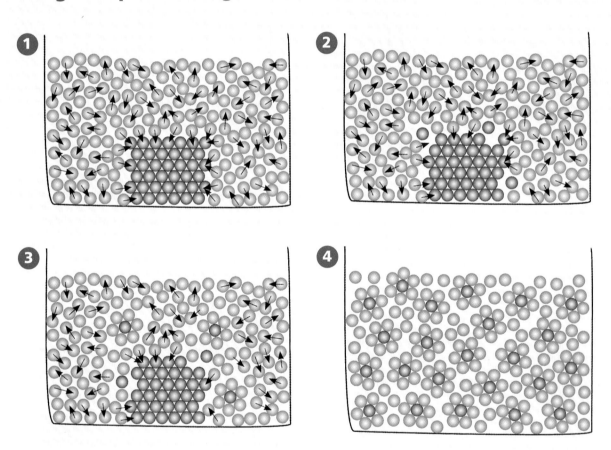

Dissolving breaks the bonds that hold particles of the solute together.

A cough drop doesn't work immediately. It needs time to dissolve and release its medicine.

The sugar particles leave the cough drop and form bonds with a few water particles. The tiny sugar-and-water groups move off into the mass of water (3).

After 20 minutes, the whole cough drop has been broken into individual particles and carried into the water. The cough drop is completely dissolved. And it happened one particle at a time (4).

> **Take Note**
>
> How could you increase the rate of dissolving of a cough drop in water? Hint: What could increase the number of collisions between the liquid and solid?

Mixtures and Solutions

When two or more materials are put together, you have a mixture. Peanuts and raisins make a good mixture for snacking. Tomato sauce with herbs poured over pasta is a mixture. A mixture of oil and vinegar is a good start for salad dressing. Any combination of materials is a mixture.

You can make a mixture of salt and water. Salt dissolves in water, just like the cough drop. After the salt has dissolved, it is no longer visible. The mixture is **transparent**. When two substances are put together and one dissolves in the other, this kind of mixture is called a solution.

Solutions have two parts. There is the part that dissolves (salt) and the part that the salt dissolves in (water). The part of the solution that dissolves is the **solute**. The part that does the dissolving is the **solvent**.

This snack of peanuts and raisins is an example of a mixture of two solids. It is not a solution.

Solutions on Earth

Remember the article called "Elements in the Universe"? In that article, we looked at the elements that make up the ocean and the atmosphere. We discovered that the ocean is mostly hydrogen and oxygen, and the atmosphere is mostly nitrogen and oxygen. As it turns out, the ocean and the atmosphere are both solutions. Let's look at the ocean and atmosphere again, this time to see how those elements are organized in solutions.

What is the largest solution on Earth? That's a tough question. Largest in volume or largest in mass? The most massive solution has to be the ocean. The ocean's depth ranges from a few centimeters to over 10 kilometers (km). And the ocean covers more than 70 percent of Earth's surface. That's a lot of sea water.

The ocean is Earth's most massive solution; that is, the largest in mass. Most of the minerals dissolved in sea water are there because of erosion.

Earth's atmosphere is the largest solution in volume. Air is considered a solution, with nitrogen gas as the solvent for other gases.

Sea water is a solution. The solvent in sea water is pure water, but sea water contains a lot of solutes. The main solute is the salt, sodium chloride, the same salt you sprinkle on food. There are thousands of other solutes in sea water, too, but in very small amounts. In fact, every element that occurs naturally on Earth can be found in sea water.

The solution that is largest in volume is Earth's atmosphere. The atmosphere covers Earth's entire surface, land and sea, and extends up about 600 km.

Earth's atmosphere is a mixture of gases called air. Air is pretty uniform in composition. It is about 78 percent nitrogen, 21 percent oxygen, and 1 percent argon, and has traces of hundreds of other substances. Does it seem a little odd to think of air as a solution? What is dissolved, and what did the dissolving? The substance that is present in greatest quantity is considered the solvent. So in air, nitrogen gas is the solvent. Oxygen, argon, carbon dioxide, water vapor, and all the other gases are solutes.

Solutions for Life

You are full of solutions. Water is the solvent for most of them. Saliva is a solution. So are sweat, urine, stomach acid, and tears. Each solution has an important function in the successful operation of a living human being.

Let's think about blood. If you spin a sample of blood in a device called a centrifuge, on top of the tube will be a clear, amber liquid called blood plasma. Plasma is a solution. The solvent is water, and the many solutes include proteins, vitamins, and minerals. There are solid parts of the blood at the bottom of the test tube. The solid portion of blood is mostly red and white blood cells. So blood is really a mixture. It is a solution with solids suspended in it.

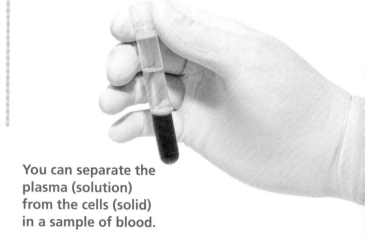

You can separate the plasma (solution) from the cells (solid) in a sample of blood.

Saliva is a solution of water and dissolved enzymes, which are chemicals that begin digestion and prevent infection.

Investigation 7: *Solutions*

Plants contain a variety of solutions in their stems, leaves, and flowers. The sweet nectar that attracts bees and butterflies is a sugar solution. The sap flowing in plant stems and leaves is a solution of sugars and salts. The solution drawn up from the roots to all the plant's cells is a solution of minerals from the soil.

In every case, water is the solvent for life. The solutes provide the raw materials that make life possible.

Solution Defined

- A solution is a kind of mixture of two (or more) substances, where one substance (solute) is dissolved in the second substance (solvent).
- In a solution, the solvent particles hit the particles of the solute and break the bonds holding the solute particles together. This is called dissolving.

Many flowering plants produce a sugary solution called nectar that attracts butterflies and other pollinators.

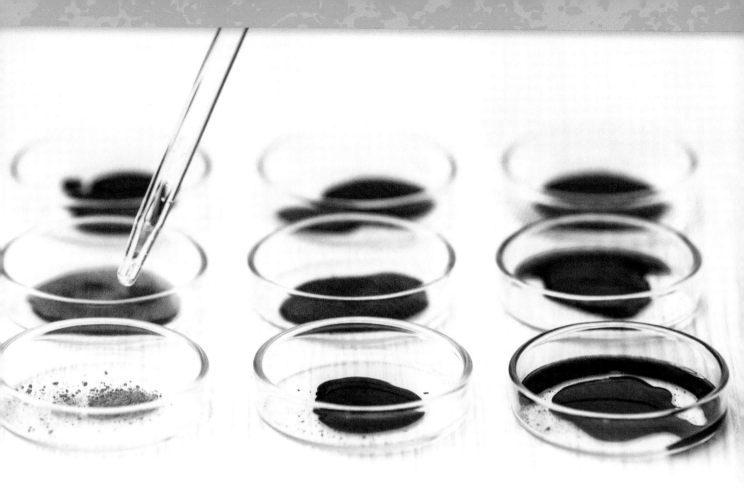

These solid dyes dissolve to form solutions when water, a solvent, is added.

- Substances that dissolve are **soluble**. Salt is soluble in water. Substances that don't dissolve are **insoluble**. Sand is insoluble in water.
- Individual solute particles are broken apart from the solid by the solvent particles. When the solute is all dissolved, the solute particles are uniformly distributed among the particles of the solvent.
- When one substance dissolves in another substance to make a solution, the particles of the two substances do not change. Solutions can be separated into their original substances. The most common way to separate a solution is by evaporating the solvent, which leaves the solute behind.

Think Questions

1. Magnesium sulfate (Epsom salt, $MgSO_4$) dissolves in water. Describe what happens at the particle level when magnesium sulfate is put into water.
2. How could a solution of magnesium sulfate and water be separated into its starting substances?
3. What are some of the solutions found in living organisms?

Investigation 7: *Solutions*

Concentration

You stop by the freezer section at the market to pick up a can of orange juice. You defrost it and try to drink the orange juice straight from the can. Yuck!

The orange juice is too thick and strong. That's because the can contains orange juice concentrate. Most of the water has been removed from the juice. The juice is too **concentrated** to drink. Once you add water to the juice, it is ready to drink.

Orange juice straight out of the orange is a solution. Water is the solvent. There are probably hundreds of solutes in orange juice, including sugars, vitamins, acids, salts, gases, and starches. The particles of all the solutes are evenly distributed among the water particles.

Orange juice is often sold in a concentrated form, because it's easy to add water at home, and less expensive to ship it as a concentrate.

At the factory, the orange juice is heated to evaporate some of the water. As water particles leave the solution, the solute particles (sugars, vitamins, and so on) stay evenly distributed in the solution, but there is less solvent. As a result, the solute particles get closer together, because there are fewer water particles between them. Here's how that works.

Look at the pot of orange juice in illustration 1. There are 100 water particles and 25 orange juice particles. The pot of juice is gently heated. Solvent (water) particles begin to evaporate, as in illustration 1a. When 75 water particles have evaporated, the orange juice looks like illustration 2.

Heating Orange Juice

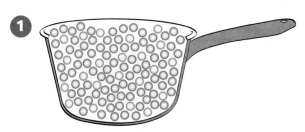

1 A representation of orange juice in a pot shows many more solvent (water) particles than solute particles.

1a Water particles evaporate when juice is heated.

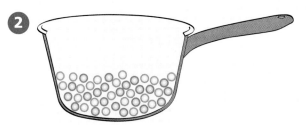

2 After much of the water has evaporated, the amounts of water and solute particles are roughly equal.

Investigation 7: Solutions

A comparison of the pot of orange juice before (1) and after (2) heating shows how much the concentration of the solution has increased.

Concentrated Solutions

Look at the two pots. What's the same and what's different?

Both pots contain the same solvent and solutes: water particles and orange juice particles. Both pots also have the same number of orange juice particles. (During heating, only water particles evaporated.)

The important difference is the amount of solvent. The fresh juice in pot 1 had 100 water particles and 25 juice particles. The evaporated juice in pot 2 has only 25 water particles and 25 juice particles.

The orange juice in pot 2 is more concentrated. That means 100 milliliters (mL) of concentrated solution in pot 2 contain more solute particles than 100 mL of solution in pot 1. The juice in pot 1 is less concentrated.

After heating and evaporation, a juice solution will have a higher concentration. To make it drinkable, we add water back in.

You can also think of **concentration** as the **ratio** of solvent particles to solute particles. There are four times as many solvent particles (water) as solute particles (juice) in pot 1. The ratio of solvent to solute is four to one. In math, that is written 4:1.

In pot 2, there are 25 solvent particles (water) and 25 solute particles (juice). The ratio of solvent to solute is 1:1. Because there is only one solvent particle for each solute particle in pot 2, that solution is more concentrated than the solution in pot 1. Pot 1 has four solvent particles for each solute particle.

Investigation 7: *Solutions* 77

Dilute Solutions

Solutions that are not concentrated are **dilute**. Dilute solutions have relatively few solute particles swimming among a lot of solvent particles.

After you add water to the concentrated orange juice, it tastes just right. Sweet and delicious. If the juice is not cold enough for your taste, you can cool it down by pouring the juice over some ice cubes (1).

The juice will flow down around the ice cubes. Energy will start to flow from the warm juice to the cold ice. The juice will start to cool (2).

Diluting Orange Juice

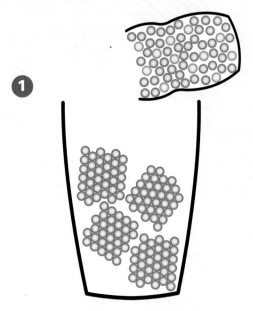

A glass of juice is poured over ice cubes.

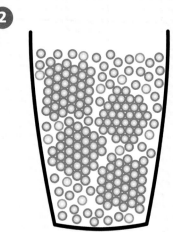

The juice flows around the ice. The juice gets cool and the ice melts.

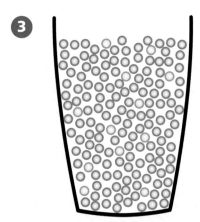

When the ice melts, the orange juice particles are distributed in a lot more water particles.

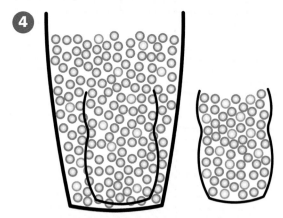

The orange juice in the large glass is less concentrated than it was before it was poured over ice.

When energy transfers from warm air to cold ice, the ice melts. The same thing happens when we drop ice into a liquid beverage: the ice melts and adds water to the drink, lowering the concentration.

But there is a catch. When energy transfers to ice, the ice melts. And when ice melts, it changes from solid to liquid. Liquid water is the solvent in orange juice. As more ice melts, there is more liquid water in the juice. By the time the ice has all melted, the juice is no longer as tasty as it was (3).

If you drop the little glass into the big glass of dilute juice, you can compare equal volumes of juice (4). Which juice sample has more juice particles? Which sample is more concentrated?

Investigation 7: *Solutions*

General Rules of Concentration

There are two ways to control the concentration of a solution. You can change the amount of solvent, or the amount of solute.

Here are the general rules of solutes.
- The more solute you use, the more concentrated the solution is.
- The less solute you use, the more dilute the solution is.

Here are the general rules of solvents.
- The more solvent you use, the more dilute the solution is.
- The less solvent you use, the more concentrated the solution is.

Parts per Thousand

Have you ever accidentally swallowed sea water? How did it taste? Salty. Very salty. If you evaporate a kilogram of sea water (a little less than a liter), you will be left with about 35 grams (g) of salt. That is about the amount of salt that will fit in the palm of your hand.

A kilogram is 1,000 g. In 1,000 g of sea water, 35 g are salt. If you think of grams as parts, then 35 parts are salt. Concentration of solutes is often reported in parts per thousand (ppt). Thus the salinity (saltiness) of the sea is 35 ppt.

Scientists have calculated the concentrations of different salts and minerals in sea water. Concentrations are measured in parts (or particles) per million, or even per billion.

Concentration of Substances in the Environment

Sea water has a high concentration of chlorine because of the salts. In every 1,000 particles of sea water, 19 of them are chlorine particles. If we look at 1 million particles of sea water (that's 1,000 times as many), 19,000 of them are chlorine. We say chlorine has a concentration of 19,000 parts per million (ppm). Reporting concentration as parts per million allows us to compare the concentration of chlorine to the concentration of other elements in sea water.

In sea water, the element magnesium is pretty concentrated at 1,290 ppm. So are sulfur (904 ppm), calcium (411 ppm), and potassium (392 ppm). Elements in lower concentrations include carbon (28 ppm), nitrogen (16 ppm), and silicon (3 ppm).

Did You Know?

You can certainly taste the salt in sea water. What about tap water? It doesn't taste salty, but there are tiny amounts of salt in the water. The concentration of the salt is so low that your taste buds cannot identify it.

Investigation 7: Solutions **81**

Most of the remaining elements are found in sea water, too, but in very low concentrations. For instance, phosphorus (0.09 ppm) is less than 1 ppm. Iron and arsenic (both 0.003 ppm) are significantly less than 1 ppm. These concentrations could be reported in parts per billion: phosphorus (90 ppb), iron (3 ppb), and arsenic (3 ppb).

Mercury

Mercury is element 80 in the periodic table. It occurs naturally in Earth's crust and in the ocean. Mercury is a toxic substance for most organisms. In humans, it can cause loss of coordination, irregular heart rate, confusion, muscle and joint pain, and in extreme cases, death. In other organisms, it can interfere with reproduction and cause mutations.

Most of the time, mercury is not a problem for living organisms. That's because the concentration is low. In sea water, mercury concentration is 0.2 ppb. That means there are only two mercury particles in every 10 billion particles of sea water.

Once mercury gets into food chains, it builds up in larger predators in a process known as biomagnification. The concentration of mercury in larger predators can reach dangerous levels for human consumption.

Problems occur when the concentration of mercury increases. The concentration of mercury in drinking water must not be greater than 2 ppb. At that amount, a person would not collect dangerous amounts of mercury.

But mercury can enter our bodies in other ways, mostly in the seafood we eat. Sardines, herring, shrimp, oysters, and clams have low mercury concentrations. These small animals grow fast. They do not live long enough to accumulate much mercury in their bodies. But larger fish eat the smaller fish and collect mercury during their lives. This is called **biomagnification**. Large fish that live a long time, like swordfish and sharks, have the most mercury (1 ppm) in their tissues. That is 500 times more concentrated than the acceptable limit in drinking water! Other fish high on the food chain, like tuna and sea bass, also have high mercury concentrations. When we eat these fish, we can accumulate too much mercury in our bodies, and our health suffers. You can check federal and local guidelines for advice on which fish to limit in your diet.

Carbon Dioxide in the Atmosphere

You have probably heard of **climate change**. Earth's climate is changing as Earth's atmosphere is slowly heating up. What does this have to do with concentration?

The concentration of carbon dioxide (CO_2) in the atmosphere is increasing. Carbon dioxide is a **greenhouse gas**. Greenhouse gases have earned this nickname because they "trap" **thermal energy** in Earth's atmosphere, heating the planet.

Carbon dioxide is produced when fossil fuels, such as gasoline or coal, are burned. We are burning more fossil fuel than ever before. So we are constantly pouring carbon dioxide into the air. At the same time, forests are being destroyed for their resources and to make room for new farms or homes. Forest plants remove carbon dioxide during photosynthesis. Less forest means less carbon dioxide is removed from the environment. More carbon dioxide is entering the air, and less is being taken out. The math is pretty simple.

Unlike smog, a visible air pollutant, carbon dioxide is invisible. A typical passenger car emits 4.7 metric tons of carbon dioxide into the atmosphere each year!

What is the concentration of carbon dioxide in the air? It varies depending on where you measure it. One of the best places to measure the carbon dioxide concentration is the top of Mauna Loa, a 4,170-meter (m) volcano on the big island of Hawaii. This volcano is in the middle of the Pacific Ocean, far from major sources of carbon dioxide and forests. The measurements are a good indicator of global carbon dioxide concentration.

When scientists began monitoring carbon dioxide in 1959, the average concentration was 316 ppm. In 2017, the concentration reached 410 ppm. In 58 years, the concentration increased 94 ppm, or 30 percent. The graph shows the rapid increase in carbon dioxide concentration. While 410 particles of carbon dioxide for every million particles of air might not seem like a lot, it is enough to change our atmosphere. Most climate scientists agree that current concentrations of carbon dioxide in the atmosphere will result in a global temperature increase of 3–4°C. That change is affecting Earth's climate, ice caps, and sea level.

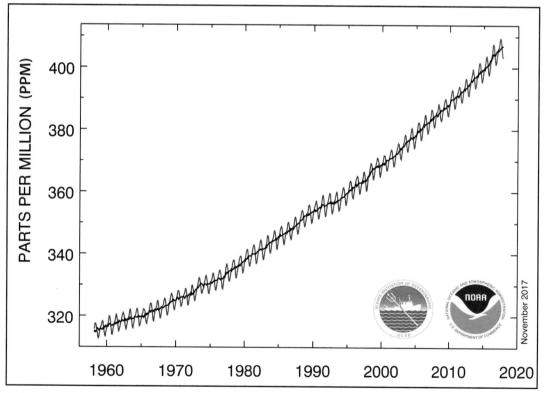

Scripps Institution of Oceanography
NOAA Earth System Research Laboratory

As the graph shows, carbon dioxide concentrations in the atmosphere are increasing at an accelerating rate. The main cause is human activity.

The ocean contains huge amounts of gold, but it is impractical to extract the metal from sea water. Nuggets like these are mined from land deposits.

Gold in the Ocean

The ocean is probably the most complex solution in the world. Just about everything you can think of is dissolved there.

One element in sea water is gold. The concentration of gold is about 0.00001 ppm. How could you get the gold out of solution? You would need something like a gold magnet.

If you did invent a gold magnet, could you get rich? Maybe. If you do the math, it looks as if you could extract 200 g of gold from every 180 billion liters (L) of sea water. How much water is that? About the amount it takes to fill 75,000 Olympic-sized swimming pools.

The gold in sea water is a little dilute. You probably want to focus on a more concentrated resource for a get-rich plan!

Think Questions

1. Why does juice taste weak after the ice in it melts?
2. What does biomagnification of mercury have to do with concentration?
3. How does the concentration of greenhouse gases relate to Earth's global climate?

The Frog Story

Scientists have noted a large decline in the worldwide frog and toad populations in the last three decades. About a third of these species are in danger of becoming **extinct**. The causes of the decline are difficult to pin down. There is evidence, however, that toxic chemicals may play a role.

Tyrone B. Hayes (1965–) is a biology professor at the University of California at Berkeley. He has studied frogs for most of his life. The first frogs he saw were in a swamp near his home in South Carolina. Now Hayes studies frogs across the planet, in Africa and North America.

Hayes found something strange going on with some of the frogs he studied. Frogs living in the wild were experiencing sex changes. The "male" frogs were producing eggs. The males were changing into females.

Hayes and his team analyzed the frogs' environment. They found traces of a common **herbicide** (herb = plant; cide = kill) in the water. The concentration was only 0.1 ppb. That's one particle in 10 billion. But tests in the lab showed that the herbicide was causing the changes.

Herbicides are designed to kill unwanted plants like weeds. Unfortunately, they can also affect other organisms in the ecosystem.

Investigation 7: Solutions

Frog populations are dropping disastrously worldwide, and dozens of species are dying off altogether. Why? The main culprits are habitat loss, deadly fungi, and toxins added to the environment by humans.

The herbicide does not kill the frogs. But it does affect the population size. How? By altering frog reproduction. The substance in the frogs' environment, even in very low concentration, makes it impossible for the frogs to produce offspring. The herbicides prevent the next generation from being born.

About 20,000 different herbicides and insecticides are used in the United States. These pesticides are designed to kill unwanted organisms, like bugs that destroy crops. But the pesticides do not all stay where they are applied. Particles of the substances are carried by wind and water to other locations. In other environments, pesticides can have unintended results. A major study released in 2017 confirmed that some pesticides used in agriculture can accidentally kill bees. Bees are essential to ecosystems for their role in pollinating plants and food crops.

The problems with pesticides are growing all over the world. Governments are working to regulate substances that might harm the environment.

Rock Solid

What does lava pouring out of a volcano have in common with a snowman?

They are both going to change phase in a short time. The liquid **lava** will **freeze** and become solid rock. The solid snowman will melt and become liquid water.

Most matter on Earth exists in one of three forms: solid, liquid, or gas. The forms are called states or **phases of matter**.

The clothes you wear, the forks and spoons you eat with, and your books and pencils are a few examples of matter in its solid phase.

The olive oil you put on your salad, the shampoo you use to wash your hair, and a refreshing glass of cold milk are examples of matter in its liquid phase.

The helium in a party balloon, the air you pump into a soccer ball, and the carbon dioxide in your exhaled breath are examples of matter in its gas phase.

Deep inside Earth, temperatures are hot enough to melt rock. This molten, liquid rock rises to Earth's surface as lava when a volcano erupts. What will happen to these glowing, flowing rivers of rock?

Investigation 8: Phase Change

Water exists naturally on Earth in all three states of matter. Snow is solid water.

Properties of the Phases of Matter

Many substances can exist in more than one phase. The snowman, for instance, is made of solid water. We have many **common names** for solid water, including ice, frost, and snow.

Water can also exist as liquid. Liquid water falls from clouds as rain and flows to your home in pipes. Earth is mostly covered by an ocean filled with liquid water.

Water also exists as gas. Water in its gas phase is called water vapor. We are usually not aware of water vapor because it is invisible. Most of the water vapor on Earth is in the atmosphere as part of the air.

Ice, liquid water, and water vapor all look different, but they are all forms of water. What is the same and what is different about ice, liquid water, and water vapor?

All three phases of water are made of exactly the same kind of particle. The chemical formula for the water particle is H_2O. Ice, liquid water, and water vapor are all made of water particles.

The thing that is different about ice, liquid water, and water vapor is the relationship between the water particles.

In the article "Three Phases of Matter," we considered how solids, liquids, and gases differ. In solids, the particles are attached to one another. The attachments are called bonds. The bonds in solids (1) are so strong that the particles cannot change positions. That's why solids have definite shape and volume.

In liquids (2), the bonds are weaker. The particles are still held close together, but they can move around and past one another. As a result, liquids flow. That's why liquids have definite volume, but their shape changes.

In gases (3), bonds do not hold the particles together. Individual particles of gas fly around in space. That's why gases do not have definite volume or shape.

Three Phases of Matter

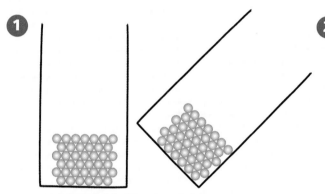

Ice in a vial will move from side to side, but will not change volume or shape.

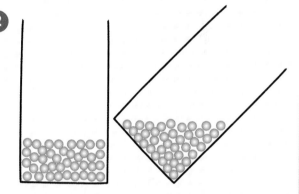

Liquid water has definite volume, but its shape changes to fit the container it is in.

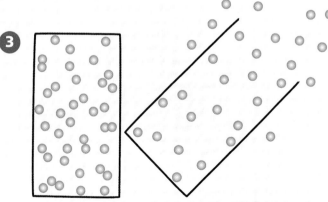

Water vapor (gas) does not have definite volume or shape. If the container is open, the gas may expand, and the particles will leave the container.

Phase Change

The snowman was not always solid. And it will not stay solid. The solid snowman will melt and turn into liquid water. The liquid lava was not always liquid. And it will not stay liquid. The liquid lava will freeze and turn to solid rock.

Change from solid to liquid and change from liquid to solid are examples of phase change. What causes substances to change phase?

Heating causes phase change. Or, more accurately, energy transfer causes phase change. Here's how it works.

Melting

When a piece of ice sits in a warm room, energy transfers from the air particles to the water particles in the ice. The kinetic energy of the water particles increases until the ice reaches 0 degrees Celsius (°C).

As more and more energy transfers to the 0°C ice, the bonds holding the water particles together start to break. When most of the bonds are broken, the water particles are no longer held in place. They start to move over and around one another.

When particles flow over and around one another, we say the substance changed from solid to liquid. The process is called melting. Substances melt when enough energy transfers to the particles of a solid to break the bonds holding the particles in place.

Solid ice cream holds its shape on top of a cone. Once it melts, it has the properties of a liquid, flowing and changing shape . . . and making a mess!

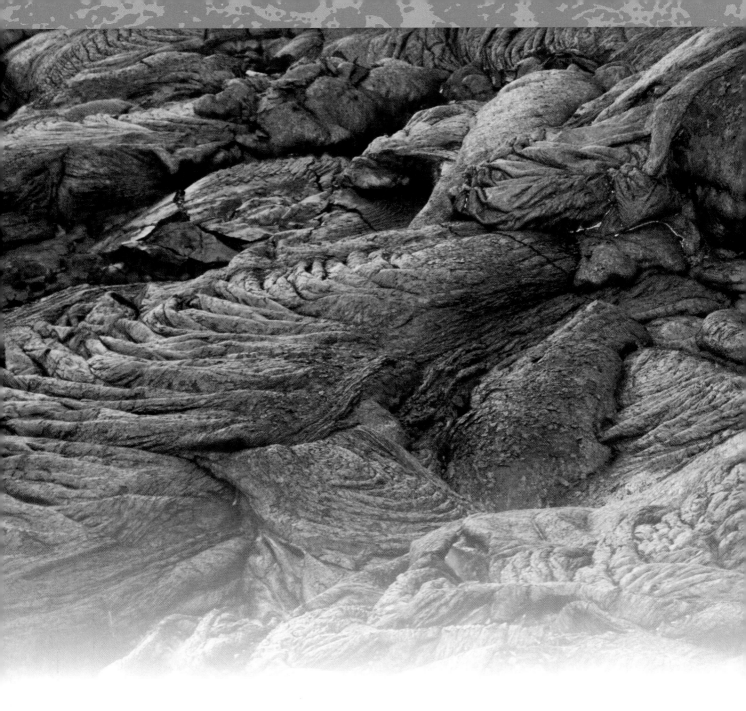

That is why the snowman melts. Energy from the Sun transfers to the water particles in the snow **crystals**. The bonds that hold particles together as a solid are broken, and the solid water changes to liquid water. The snowman changes into a hat and scarf on top of a puddle of water.

Freezing

What about the lava? How does it change phase? When lava pours out on Earth's surface, it is extremely hot (up to 1,100°C). The kinetic energy of the rock particles is so great that most of the bonds holding them together have been broken. The rock particles move over and around one another. The rock is liquid, so it flows down the side of the volcano.

Hot, liquid lava will freeze and change to solid rock.

Air is cooler than lava . . . a lot cooler. Energy from the liquid rock particles transfers to the air particles. The rock particles lose kinetic energy, and the mass of lava cools. As the lava cools, stronger bonds form between the rock particles. When enough energy has transferred from the rock particles, the particles are locked in place by the bonds.

When particles stop flowing over and around one another, we say the substance changed from liquid to solid. The process is called freezing. Substances freeze when enough energy transfers away from the particles of a liquid.

That's why the liquid lava freezes and becomes solid rock. Energy transfers away from the rock particles, bonds hold the particles together in a fixed position, and the rock changes from liquid to solid.

When warm water vapor in the air comes into contact with a cold window, energy transfers from the gas to the glass. As the gas loses energy, invisible water vapor changes to tiny liquid water droplets that we can see.

Evaporation

Let's get back to the snowman. After a day or two, all that remains is the hat and scarf. Even the puddle of liquid water has disappeared. Where did the water go?

As sunshine falls on the puddle of liquid water, energy transfers to the water particles. The kinetic energy of the particles increases. When enough energy transfers to a particle, the particle breaks all the bonds holding it to the mass of liquid. The particle breaks free and flies into space. The water changes phase again, but this time from liquid to gas.

The phase change from liquid to gas is called **evaporation** (or vaporization). Water in the gas phase is called water vapor. The individual water particles are too small to see, so water vapor is invisible. Water vapor enters the air and becomes part of Earth's atmosphere.

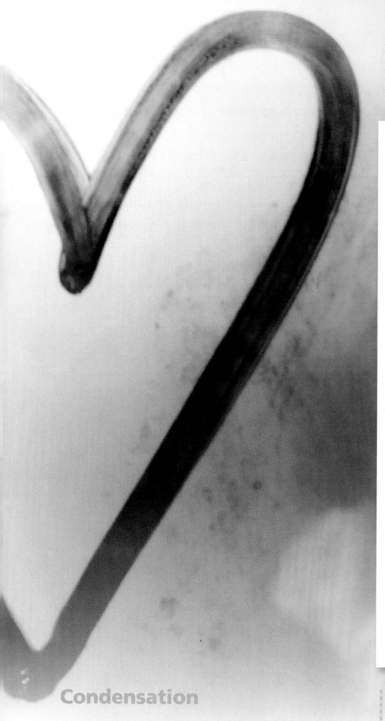

Evaporation and Condensation

Liquid water evaporates when it is heated, forming water vapor (gas). Water vapor condenses back to a liquid on the outside of a cup containing ice.

Condensation

Water can change from gas to liquid, too. The process involves energy transfer. Can you predict what energy transfer takes place?

When energy transfers away from the water vapor particles, they lose kinetic energy. When enough energy has transferred from the particles, bonds form between them. The water changes phase from gas to liquid. The process is called **condensation**. Substances condense when energy transfers away from the particles of a gas.

Look at the illustration of an experiment. A pan of liquid water is heated. Water evaporates. The water vapor condenses on a cup filled with ice. Study how the water particles change phase from liquid to gas, and then back to liquid. You should be able to see where evaporation and condensation are taking place.

Investigation 8: Phase Change

Temperature at Phase Change

There are two important things to understand about melting and freezing. **Substances do not have to be cold to freeze.** *Freeze* just means changing phase from liquid to solid. This happens at different temperatures for different substances. Granite freezes at about 1,650°C. Oxygen freezes at –218°C. Every substance has its own freezing temperature. Any solid substance is technically frozen, even a hot metal pan on the stove.

Freezing temperature = melting temperature. A substance freezes and melts at the same temperature. Water, for instance, has a **freezing point** and **melting point** of 0°C. If you move a piece of ice from a freezer to a warm room, the ice will warm up until it reaches 0°C. Then it will melt. If you put a cup of warm water in a freezer, the water will cool until it gets to 0°C. Then it will freeze.

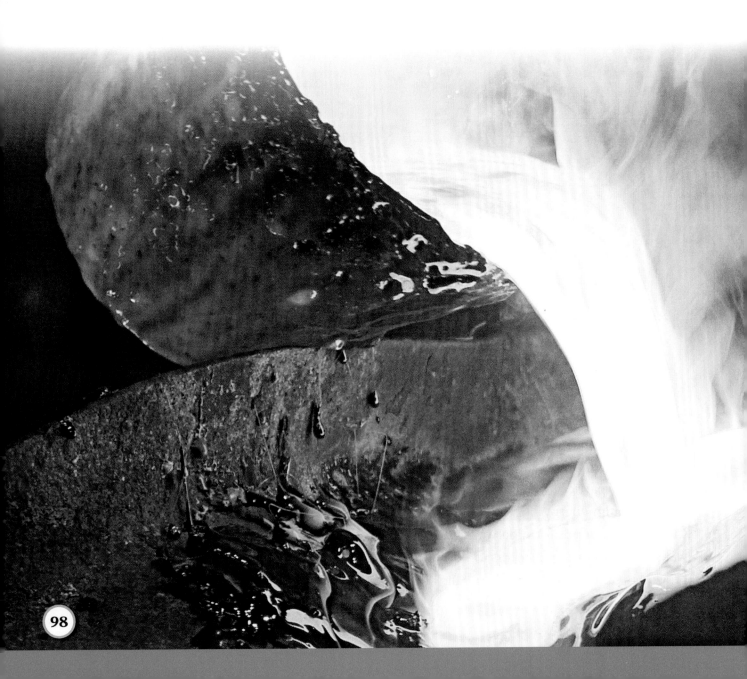

Evaporation temperature = condensation temperature. The temperature at which a substance evaporates is the same as the temperature at which it condenses. Water, for instance, evaporates and condenses at 100°C. It all depends whether you are heating it up or cooling it down.

Particle Relationships

The phase of a substance is determined by what is happening between the particles in the substance. Particles in solids have strong bonds, particles in liquids have weak bonds, and particles in gases have no bonds.

Phase Change Temperatures

Substance	Freeze/melt (°C)	Condense/evaporate (°C)
Helium	–272	–269
Oxygen	–218	–183
Nitrogen	–210	–198
Carbon dioxide	—	–78
Chlorine	–101	–34
Mercury	–39	357
Water	0	100
Sodium	98	883
Lead	327	1,749
Aluminum	660	2,519
Calcium chloride	775	1,936
Sodium chloride	801	1,465
Silver	962	2,162
Gold	1,064	2,856
Copper	1,085	2,562
Iron	1,538	2,861
Tungsten	3,422	5,555

The freeze/melt temperatures and condense/evaporate temperatures for some common substances are shown here. Carbon dioxide is a special substance. Do you see that it has no freezing and melting point? Keep reading to find out why.

In the refining process, gold is heated to its melting point (1,064°C) to remove impurities. Melted gold can then be poured into molds.

Investigation 8: Phase Change

Phase-Change Vocabulary

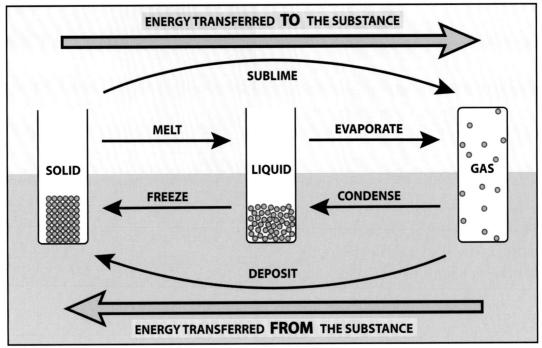

It's all about energy transfer. Adding energy to a substance can cause melting, evaporation, or sublimation. Removing energy from a substance can cause condensation, freezing, or deposition.

This illustration summarizes how energy transfer affects phase change. The top half shows how substances go from solid to liquid to gas as energy transfers *to* the particles of the substance. The bottom half shows how substances go *from* gas to liquid to solid as energy transfers from the particles of the substance.

Notice that a substance can go straight from solid to gas. This process is called **sublimation**. Carbon dioxide is a substance that sublimes. And when energy transfers the other way, carbon dioxide **deposits**. It changes from a gas to a solid without going through a liquid phase. That's why solid carbon dioxide is called **dry ice**.

Think Questions

1. What causes a substance to change from one phase to another?
2. Why is the freezing temperature for a substance the same as the melting temperature?
3. What happens to water particles as a cup of ice melts and then evaporates?

Heat of Fusion

It's a warm day, and you are at a party or a picnic. You are thirsty, so you go to the ice chest and take out an ice-cold beverage. You pop the top and take a sip. So cold and refreshing!

The ice chest is a great invention. It is simple to maintain, easy to use, and very efficient for cooling drinks. But did you ever think about how it works? It is not quite as simple as it seems.

Remember that energy transfers from warmer objects to cooler objects. Can you figure out the energy transfers required for these cans to stay cool in ice?

Investigation 8: *Phase Change* **101**

How Does Ice Make Things Cold?

If you ask your little brother how an ice chest works, he might have this idea. "Ice is cold. When you put it in an ice chest, it makes the whole inside of the chest cold. When you put drinks in with the ice, the cold goes into the drinks. Cold is stronger than hot, so it just takes over. That's why drinks get cold in an ice chest."

That's an interesting explanation, but it is not true. We know that cold is not a substance, and it cannot transfer to the warm drink. *Cold* just means that particles have low energy. The lower the energy of particles, the cooler the material. Objects become cooler when energy transfers from them to another location. Drinks in an ice chest get cold because energy transfers from the drinks (particles with higher energy) to something else (particles with lower energy). The "something else" in the ice chest is the ice itself. And what happens to the ice? It melts.

Iced tea stays chilled because energy transfers from the liquid tea to the solid ice cubes. The transferred energy causes the ice to melt.

What Really Happens When Ice Melts?

Boiling water is 100 degrees Celsius (°C). Ice that has just formed is 0°C. Ice that has just melted is 0°C, too. If we mix 100 grams (g) of 100°C water with 100 g of 0°C water, in a moment the mixture will reach equilibrium. We will have 200 g of water at 50°C.

The equation that lets us predict the final temperature when we mix equal volumes of water of two different temperatures looks like this.

$$T_f = \frac{T_h + T_c}{2}$$

T_f = final temperature
T_h = temperature of hot water
T_c = temperature of cold water

When we add 0°C ice cubes to 100°C water, the final temperature is not halfway between the two. Read to find out why the temperatures don't meet in the middle.

We can use the equation to predict that the final temperature of the mixture described on the previous page will be 50°C. The final temperature is also known as the equilibrium temperature.

$$T_f = \frac{T_h + T_c}{2} = \frac{100°C + 0°C}{2} = 50°C$$

Now let's mix 100 g of 100°C water with 100 g of 0°C ice. The starting temperatures are 100°C and 0°C. So we might predict that the temperature of the mixture at equilibrium will again be 50°C. But it is not. The equilibrium temperature is only 10°C.

Mixing Hot Water and Ice

Energy transfers from particles in the hot water to particles in the ice.

It takes a lot of energy to melt ice. Ice is the solid phase of water. Water particles in ice are held in place by forces called bonds. In order for solid water to turn into liquid water, the bonds must be broken. It takes energy to break bonds. That's where most of the energy transferred from the hot water goes. The energy breaks the bonds between particles and changes water from a solid to a liquid.

But the energy that melts the ice does not change its temperature. Ice is 0°C, and the liquid water it turns into is also 0°C. The energy that transfers to ice does not change the kinetic energy of the water particles. It just breaks bonds. The energy that breaks bonds to change solid water into liquid water is called **heat of fusion**.

At 0°C, the ice on a lake can melt. Heat of fusion breaks the bonds holding ice particles together. But the water and ice temperatures stay at 0°C until all the solid turns to liquid.

Investigation 8: *Phase Change*

Calculating Heat of Fusion

We can calculate the heat of fusion using the equation for calculating **calories**. We know that the hot water went from 100°C to 10°C. That is a change of 90°C. Let's use that to calculate the energy transferred from the hot water. We will refer to this transferred energy as cal_h.

$cal_h = \Delta T \times m$
ΔT = temperature change
m = mass
$cal_h = 90°C \times 100\ g = 9{,}000\ cal$

We also know that the ice went from 0°C to 10°C. That is a change of 10°C. Let's use that to calculate the energy transferred to the ice (cal_c).

$cal_c = \Delta T \times m$
$cal_c = 10°C \times 100\ g = 1{,}000\ cal$

Snow doesn't immediately turn to water on the first warm day. It takes a lot of energy—the heat of fusion—to change a substance into a different state of matter.

It appears as though the number of calories transferred from the hot water exceeds the number of calories transferred to the cold ice by 8,000 calories. But we know that energy is conserved. No energy is ever created, destroyed, or lost during energy transfers. So what happened to the 8,000 calories?

The 8,000 calories were used to break the bonds to melt the 100 g of ice. We can do one last calculation to find out how many calories are needed to melt just 1 g of ice. Divide the total energy by the number of grams of ice in the sample.

8,000 cal/100 g = 80 cal/g

The heat of fusion for water is 80 cal/g. That means for every gram of ice that melts, 80 calories of energy transferred from someplace to make that happen.

Investigation 8: *Phase Change*

Using Energy Transfer to Make Things Cold

Heat of fusion is what makes the ice chest so good at cooling drinks. The best way to set up an ice chest is to fill it about half way with crushed ice. Then pour in just enough water to float the ice. That will maximize the area of contact where energy can transfer out of the drink containers through conduction. The water will transfer energy to the ice, and some of it will melt. But as soon as the water gets down to 0°C, ice will stop melting. Why? Because the kinetic energy of the particles in the 0°C water will be the same as the kinetic energy of the particles in the 0°C ice. There will be no more net energy transfer. The system is in equilibrium at 0°C.

In go the room-temperature drinks. They sink down into the ice and water. Energy starts to transfer from the surface of the 20°C cans to the 0°C water surrounding them. As a can transfers kinetic energy to the water, the drink inside the can starts to transfer energy to the can material. The water warms up as the can and the drink inside cool down.

Energy transfers from warmer bottles to cold ice in an ice chest. Eventually, the ice-chest system will be in equilibrium, with everything at the same temperature.

Energy Transfers in an Ice Chest

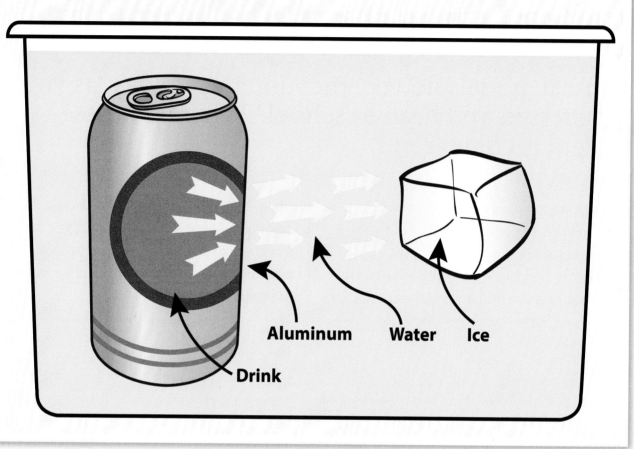

Energy transfers from drink particles to aluminum particles. The drink cools down. Energy transfers from the aluminum particles to water particles. The aluminum can cools down. Energy transfers from water particles to ice particles. Water cools down and ice melts.

But the water does not stay warm. It transfers energy to the ice. Ice melts, keeping the temperature of the water at 0°C. Energy continues to transfer from the drinks to the ice until everything in the ice chest is at 0°C.

What has changed? Energy has moved from the drinks to the ice. The drinks are cold, and some of the ice has melted. The ice-chest system uses kinetic energy from the drinks to melt ice. The result is cold drinks. The ice chest uses energy transfer to cool the drinks.

Think Questions

1. What is heat of fusion?
2. What happens at the particle level when you put ice cubes in a glass of room-temperature lemonade?

Better Living through Chemistry

What materials do you encounter every day, as you get dressed and head to school? Where do these materials come from?

Plastic, nylon, and glass—you won't find these on the periodic table. Yet these substances are part of our daily lives. Substances like plastic are called **synthetic** materials, because humans synthesize (make) them. Materials scientists and chemical engineers figure out how to combine and rearrange **atoms** of substances to create a new substance that is stronger, cheaper, safer, or otherwise more useful.

But you can't make something from nothing. Where do these scientists and engineers get their ingredients? Earth's natural resources provide the raw materials for everything you see around you, from cell phones to medicines to rockets.

Starting Your Day

Let's start right at the beginning of the day. Your phone alarm beeps, and you silence it, rubbing your sleepy eyes. What's the phone made of?

The screen is made of glass that is mostly silicon and oxygen. Humans first made glass by heating crushed sand or a mineral called quartz with other substances. These natural resources are still used to make some modern kinds of glass.

Many different natural resources were used to synthesize the materials in these colorful wardrobe accessories.

Investigation 9: *Reaction*

Think petroleum is just for gasoline to fuel vehicles? Think again! These plastic containers are petroleum based. Very often, so are their personal-care contents.

The circuitry of the phone relies on several metals, including copper, gallium, gold, indium, magnesium, palladium, platinum, silver, tin, and tungsten. These elements are extracted from rocks and ores. After processing, they end up in your phone.

Petroleum Products

You crawl out of bed and make your way to the bathroom. On the shelves are dozens of products to help you get ready for your day. What are they made of?

Shampoo, hand lotion, nail polish, and sunscreen all have something in common—they include ingredients made from **petroleum**. The bottles holding these substances are usually made of plastic, which is also made from petroleum. Hairbrushes, combs, toothbrushes, hair dryers, shower curtains, and even the toilet seat are typically made of plastic. That's right, petroleum again!

What is petroleum? It's not an element, so it's not on the periodic table. The raw material for petroleum came from dead plants and animals. Their remains were buried hundreds of millions of years ago. Instead of decaying, these remains changed into petroleum, which is made mostly of carbon and hydrogen. It is also called **crude oil**.

This natural resource is mined or collected. It is a nonrenewable resource because it was created by geological processes that take many millions of years. A refinery separates the crude oil into different **products**. Each product is a different combination of carbon and hydrogen. Most of the refined products are used for fuel. Fuels include gasoline, which powers cars, trucks, and buses. Other refined products include oils, waxes, and raw materials for plastics and medicines.

Crude oil is extracted from the ground using special machinery.

When materials scientists and chemical engineers figured out how to create plastics from petroleum, they changed the way we live. Instead of packaging shampoo in a glass bottle, how about using plastic? If it slips out of your wet hand in the shower, there won't be any broken glass to hurt your feet. Plastic means that food can easily be packaged for eating on the go, using plastic utensils. Plastic means that hospitals can store clean equipment in sealed packages, reducing the risk of infection.

Medications

During allergy season, you might take a pill to clear your sinuses. What are the ingredients in this synthetic substance? Many medications are **compounds** built from different amounts of the elements carbon, hydrogen, and oxygen, with a few other kinds of atoms mixed in. Biochemical engineers use **chemical reactions** to create these new substances. Medical researchers test new substances and help the engineers decide what to create next.

Most of the medicines we take are engineered from the six most common elements in the body: hydrogen, oxygen, carbon, nitrogen, phosphorus, and sulfur. Different combinations of these elements form compounds that have dramatically different effects in the body.

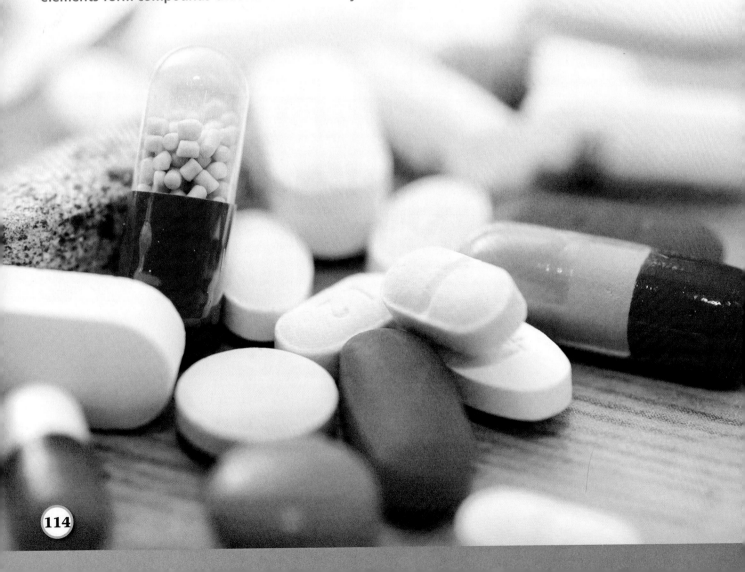

Clothing makers choose synthetic fabrics and fibers for many reasons. Some resist wrinkles, hold colors, won't shrink, or are machine washable. Others are waterproof or warm. Still others are stretchy, soft, or static-free.

Getting Dressed

You put on your clothes for the day. Your T-shirt is 100 percent cotton. But your other clothes are made with petroleum-based materials, such as nylon and polyester. Rayon is synthetic, too, made from plants. Not only do shirts, blouses, pants, and skirts have synthetic fabrics, but also sweaters, windbreakers, rain coats, sneakers, sandals, flip-flops, and tote bags. Then there are nylon umbrellas, zippers, and hangers. It is hard to believe that all these products are made from materials that come from the ground. But that is true for all petroleum-based products.

Did You Know?

Plastic does not biodegrade (decompose in nature) quickly. However, it can break into smaller pieces that may be washed into the ocean and cause problems in the ecosystem.

The kitchen is full of synthetic materials—even some of the food!

The Kitchen

You go into the kitchen for breakfast. You are surrounded by more petroleum products. They include nonstick pans, plastic bottles, drinking straws, plastic-foam egg cartons, trash bags, milk jugs, waxed paper, and lunch boxes. Even tablecloths, mops, sponges, and dish scrubbers can be petroleum-based products. The windows in the kitchen are made of glass, as is the glass you fill with juice. The stainless-steel spoon you use to eat your cereal? It's a manufactured metal, combining raw iron, carbon, and chromium.

At least your food doesn't contain synthetic materials, right? Not so fast. Many prepared foods contain synthetic vitamins, colors, and flavors. Folic acid is an important B vitamin. It is naturally in some foods, but people often consume less than the recommended amount. Without enough folic acid, pregnant women are much more likely to have a child with birth defects. Now the United States Food and Drug Administration requires food manufacturers to put extra folic acid into breads, cereals, flours, pasta, and other grain products. This synthetic vitamin helps prevent birth defects of the brain and spinal cord.

Heading to School

You reach for your sunglasses. Yes, those are also likely to be plastic. If you are not bicycling or walking, you are probably going to school in a vehicle. The car or bus is powered by petroleum-based fuel (unless you're in an electric vehicle charged by solar panels). You may be riding on asphalt streets. Asphalt is made with tar, which is another petroleum product.

Phew! You made it to school after a busy morning. Look around the classroom. What synthetic materials are here?

Synthetic materials are a part of your daily life.

Think Questions

1. What are some benefits of using synthetic materials?
2. What are some drawbacks of using synthetic materials?
3. Why are reusable grocery bags becoming more popular?

Investigation 9: *Reaction* **117**

Blowing into limewater turns the clear liquid white.

How Do Atoms Rearrange?

The cookie dough is wet and gooey when it goes into the oven, but soft and crusty when the baked cookies come out. The atoms in the dough rearranged during a chemical reaction and resulted in a delicious new substance.

When you blow air through a straw into a cup of limewater, something happens. After a couple of breaths, the limewater is a little hazy. After a couple more breaths, the limewater is definitely cloudy. After a dozen breaths, the cup of limewater is as white as milk.

Cooking often involves chemical reactions. The cookie is made of new substances produced when substances in the dough react as it bakes.

What turns the limewater white? Plain air bubbled through limewater does not turn limewater white. Something in your breath does it: carbon dioxide.

Carbon dioxide is not white. It doesn't just show up in the limewater. What happens is that carbon dioxide interacts with something in the limewater to form the white substance. The white substance is not carbon dioxide or limewater. It is a new substance.

This process is called a chemical reaction. During a chemical reaction, starting substances (like carbon dioxide and limewater) change into new substances (like the white material). Starting substances are called **reactants**. New substances are called products.

Take Note

What are the chemical reactions you have observed in class so far? What were their reactants and products?

Investigation 9: *Reaction*

Figuring Out New Products

Carbon dioxide is one of the reactants in this reaction. Calcium hydroxide is the other reactant. Limewater is the common name for a solution of calcium hydroxide and water. You can set up a **chemical equation** to help figure out what the white product might be. The chemical formula for carbon dioxide is CO_2. The formula for calcium hydroxide is $Ca(OH)_2$. The reactant side of the equation looks like this.

$$CO_2 + Ca(OH)_2 \rightarrow$$

This is how you read the equation: "One particle of carbon dioxide and one particle of calcium hydroxide react to yield . . ."

What could the products be? Remember that the smallest particle of an element is an atom. You can use atom tiles to represent a particle of a compound such as carbon dioxide. Make representations of a carbon dioxide **molecule** and a molecule of calcium hydroxide. You can then move the atoms around to model a chemical reaction and make products. An atom-tile representation of the equation is shown below.

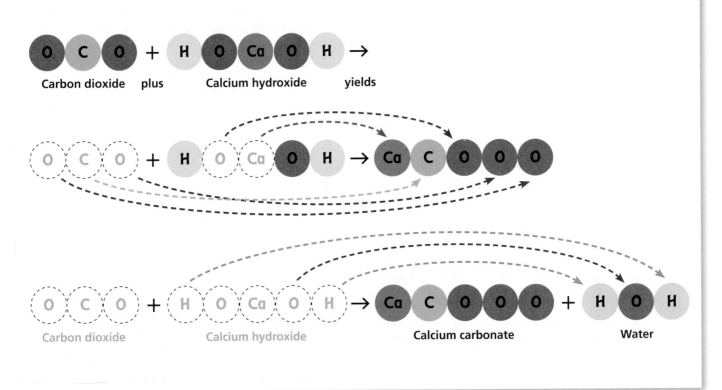

Limewater Reaction

In a chemical reaction, bonds between atoms are broken and new bonds form. But the number and kind of atoms stays the same, no matter how they are arranged.

The test tubes show the limewater investigation before, during, and after the chemical reaction. The substances change, but the total amount of matter in the system does not.

The white substance that formed in the reaction did not dissolve in water. That's a clue. Check to see what insoluble substance you can make by rearranging the atoms in the reactants. Remember, calcium carbonate ($CaCO_3$) does not dissolve in water. The atoms needed to make a calcium carbonate particle are in the reactants. You can move those atoms to the product side of the equation to produce calcium carbonate.

What's left on the reactant side of the equation? Two hydrogen atoms and one oxygen atom. These combine to form a particle of water (H_2O) on the product side.

Notice that after the reaction has happened, the reactants are gone. They no longer exist. All the atoms that were in the reactants are now in the products. And the number of atoms on the product side is exactly the same as the number of atoms that started on the reactant side. During reactions, no atoms are destroyed or created. Matter is conserved. That means the number of atoms must always be the same on both sides of the equation.

Written with chemical formulas, the limewater reaction looks like this.

$$CO_2 + Ca(OH)_2 \rightarrow CaCO_3 \downarrow + H_2O$$

The down arrow next to the formula for calcium carbonate means the substance is a **precipitate**. A precipitate is a solid product that does not dissolve in the liquid product.

Combustion Reactions

Have you ever seen a blimp slowly sailing overhead? It is an interesting vehicle. How can it fly with no wings? It stays aloft by being lighter than air. It floats in the air like a cork floats in water: it is less dense than the material around it. To make the blimp lighter than air, it is filled with helium, element 2. A blimp is a huge, motorized, helium balloon.

The huge airship *Hindenburg* was designed to be filled with helium. Because of wartime restrictions, it contained the highly flammable gas hydrogen instead, with fiery results.

During the 1930s, lighter-than-air craft like blimps were developed for transportation. One of the largest such crafts ever built was the *Hindenburg*. It was 245 meters (m) long and filled with hydrogen. Hydrogen, like helium, is much lighter than the air in our atmosphere.

The *Hindenburg* was the largest structure ever to fly. But on May 6, 1937, this huge structure caught fire. In 37 seconds, the entire aircraft was destroyed.

At this time, transatlantic passenger flights were rare, so news crews were waiting at the landing site to document the arrival. They didn't know they would be documenting a tragedy.

The main cause of this catastrophe was the hydrogen that allowed the *Hindenburg* to fly in the first place. Hydrogen is extremely flammable, and the aircraft was filled with it. Once it caught fire, it burned rapidly. **Burning** is a reaction called **combustion**. The reactants were the hydrogen gas (H_2) in the *Hindenburg* and the oxygen (O_2) in the air. You can use atom tiles to see what products formed.

H H + O O → H O H

In just over half a minute, the *Hindenburg* burst into flames and was incinerated. As the blimp plunged to the ground, many survivors jumped out of windows and managed to run away.

The main force lifting the space shuttle off the launch pad comes from a combustion reaction. The reaction releases a powerful push of gases that thrusts the spacecraft upward.

Water! When hydrogen burns, the single product is water. But look, something is wrong with the equation. One oxygen atom is left over. How can this be fixed? The solution is to react two hydrogen particles with one oxygen particle to form two particles of water. Here is the balanced equation.

Take Note

Methane (CH_4) is the main ingredient in natural gas. The products that form when methane burns are carbon dioxide and water. Write a balanced equation showing the combustion reaction for methane and oxygen.

H H + H H + O O → H O H + H O H

Investigation 9: Reaction

Aluminum foil and copper chloride react to form copper metal and aluminum chloride solution.

Metal Reactions

Modern chemistry grew out of ancient experimental work called **alchemy**. One of the goals of the early alchemists was to turn common metals into gold. They never succeeded.

But changing one metal into another with a simple reaction is possible in some cases. For example, drop some aluminum foil (Al) into a solution of copper chloride ($CuCl_2$). A reaction takes place.

When the reaction is complete, the silvery aluminum is gone. In its place is a reddish-brown precipitate of copper metal (Cu) and a solution of aluminum chloride ($AlCl_3$). This is the reaction equation.

$$CuCl_2 + Al \rightarrow AlCl_3 + Cu$$

A color change is one sign of a chemical reaction. The thin layer of copper covering the Statue of Liberty has reacted with substances in the air to form blue-green copper compounds.

There is a problem. Do you see it? The chlorine atoms are not balanced, so it is not possible for one particle of each reactant to react. If you start with two $CuCl_2$ particles, then you have four chlorine atoms. That's too many. To balance the equation, you need to start with three $CuCl_2$ particles and two Al particles. The balanced equation looks like this.

$$3CuCl_2 + 2Al \rightarrow 2AlCl_3 + 3Cu$$

So did you change one element into another element? No, the chemical reaction changed a solution and a solid into a different solution and a different solid. All matter was conserved.

Summary

The particles of all substances are made of atoms. An atom is the smallest particle of an element. The kind, number, and arrangement of atoms determine the kind of substance. Chemical reactions create new substances.

During reactions, atoms rearrange. The atoms in the particles of the reactants rearrange to form the particles of the products.

Equations describe reactions. Atom representations or chemical formulas can record the substances in equations.

Batteries use chemical reactions to transfer chemical potential energy to electrical energy.

Reaction equations must balance. The number of atoms of each kind must be equal on both sides of the equation. Balance is achieved by changing the number of particles reacting.

Matter is conserved. Particles of substances are created and destroyed during chemical reactions. Atoms are not created or destroyed during chemical reactions. They just rearrange to create particles of new substances.

Think Questions

1. What is destroyed and what is created during chemical reactions?
2. Write the equation for the reaction between hydrogen and oxygen. Use chemical formulas for the substances.
3. The reaction of the mystery mixture with water produces a gas. How could you test to see if the gas is carbon dioxide?

Fireworks

Firework shows are carefully planned. They feature explosions of light and sound, often timed to music. But how does it all happen? Fireworks are wonderful chemical reactions!

The earliest known use of fireworks was in 7th-century China. They were used to scare away enemies. Fireworks also appeared at important events and festivals. They are still seen at the Spring Festival (Chinese New Year).

Over the centuries, more and more countries traded with China. The use of fireworks spread. Scientists and engineers developed new ways to launch fireworks and to add color. They packaged the reactants in different ways. These changes produced multiple explosions and different shapes.

What special chemical reactions allow color to sparkle in the night sky? The answer lies in combustion.

Combustion

A combustion reaction happens every time something burns. A campfire is an example of combustion. The wood burns in the presence of oxygen (O_2). It transfers a lot of energy in the form of thermal energy and light. Here is a general equation for a campfire.

$$(\text{wood}) + O_2 \rightarrow CO_2 + H_2O + \text{heat/light}$$

This equation does not show the amount of each substance, or the chemical formula for wood. It also does not show products such as ash and smoke. But it does show an essential reactant, oxygen.

Combustion is like any reaction. Atoms from the reactants rearrange to form the products. But in combustion, the rearrangement results in energy transfer to the environment. This is called an **exothermic reaction**. The energy can be in the form of light or thermal energy.

Fireworks are specially designed combustion reactions. The display of lights in the night sky is caused by many exothermic reactions. Firework designers fit all the reactants into a small package that is launched into the air.

When a lit fuse ignites black powder fuel, a combustion reaction sends a fireworks tube skyrocketing into the air. At the top of the flight, another combustion reaction produces the light, color, and sound spectacle.

Substances

Pyrotechnic engineers are always making fireworks bigger, brighter, and more colorful. **Pyrotechnics** is the science that produces these shows. Pyrotechnic devices have five basic ingredients.
- Fuel
- Oxygen source
- Chemicals that produce color when burned
- Chlorine to intensify the color
- A binder to hold it all together

Oxidizers provide oxygen for the combustion reaction. They often contribute colors as well. The oxidizer is a key part of combustion. It breaks the bonds in the fuel. When bonds break, they release stored energy.

Elements used to create color for fireworks

Element number	Symbol	Metal	Effect
3	Li	Lithium	Medium red
4	Be	Beryllium	White
11	Na	Sodium	Yellow
12	Mg	Magnesium	Intense white and sparks
13	Al	Aluminum	Silver and white flames
19	K	Potassium	Pinkish violet and sparks
20	Ca	Calcium	Orange
22	Ti	Titanium	White and silver sparks
26	Fe	Iron	Red to yellow sparks (gold)
29	Cu	Copper	Blue to blue-green
30	Zn	Zinc	Smoke
37	Rb	Rubidium	Violet-red
38	Sr	Strontium	Intense red
51	Sb	Antimony	Glitter effects
55	Cs	Cesium	Sky blue
56	Ba	Barium	Apple green

Elements used as fuel or oxidizers for fireworks

Element number	Symbol	Name	Use
6	C	Carbon	Fuel and propellant
8	O	Oxygen	Oxidizer
15	P	Phosphorus	Fuel
16	S	Sulfur	Fuel
17	Cl	Chlorine	Oxidizer
19	K	Potassium	Oxidizer
37	Rb	Rubidium	Oxidizer
55	Cs	Cesium	Oxidizer

Many of the elements (outlined in pink) used to create the colorful effects of fireworks are metals. Other elements (outlined in dark blue) are the fuel, or oxidizers, for fireworks.

Safety

Building and lighting fireworks is dangerous. Careless use of even small firecrackers can injure people. Pyrotechnicians are trained professionals. They design, build, and conduct large public displays. They have studied the properties of the substances in fireworks. They handle fireworks with care to protect the public and themselves.

Firework combustion produces smoke and dust particles. These products contain the atoms of the reactants. Many of the metals in fireworks are toxic. Researchers are working to develop "green" fireworks. They use substances that are less toxic and may cause less environmental damage.

The next time you see a wonderful fireworks display, stop to think about it. What substances created those colors? How were they packaged for combustion and explosion? What could be the **limiting factors** for the combustion reaction? Who is the pyrotechnician running this show? Now that you know more about fireworks, they may seem even more beautiful.

Think Questions

1. What are some criteria and constraints that pyrotechnic engineers must consider?
2. How does combustion produce thermal energy and light?

Investigation 9: Reaction

Antoine-Laurent Lavoisier: The Father of Modern Chemistry

When you gaze closely into the flame of a burning candle, you are watching a combustion reaction before your eyes.

Before scientists understood combustion, the flame was mysterious, leading to more questions and hypotheses. One man's careful experimentation helped answer these questions. Antoine-Laurent Lavoisier (1743–1794) was born into a wealthy French family. As a young man, he practiced law and owned a tax-collecting agency. Lavoisier was an important citizen in his community.

But Lavoisier is remembered for his passion for science and his hard work. Early in the morning and late at night, he studied mathematics, geology, physics, biology, and chemistry. His work in geology was remarkable. It led to his election to the French Academy of Sciences when he was only 25 years old.

Antoine-Laurent Lavoisier, one of the best-known French scientists of his day, is considered the father of modern chemistry.

In the 18th century, chemical reactions were still a bit of a mystery. Then, along came Lavoisier, the first scientist to conduct truly quantitative chemistry experiments.

Studying Combustion

Lavoisier's interests turned more and more toward chemistry. One of many things he studied was combustion, the chemical reaction we usually call burning. During the 1600s and 1700s, chemists incorrectly believed that all flammable substances contain an odorless, colorless substance called phlogiston. When a substance burned, they thought it gave up this phlogiston. The material left behind weighed less than it did before.

Lavoisier thought the theory of phlogiston was wrong. He called it a "fatal error to chemistry." He set out to disprove the phlogiston theory.

In 1772, Lavoisier performed a series of experiments. He carefully weighed materials before and after he burned them. His results showed that substances such as sulfur and phosphorus actually gain weight when burned.

Lavoisier transformed science by his insistence on accurate measurement.

Priestley's Experiment

In 1774, Lavoisier met the English chemist Joseph Priestley (1733–1804). Priestley said that he had discovered a "new kind of air" by burning mercuric oxide (HgO). Priestley tested animals in a closed container with this new air. He found that they lived longer than animals in closed containers with regular air.

Lavoisier repeated Priestley's mercuric oxide experiments. Unlike Priestley, he carefully recorded data on all parts of his experiments. He collected the gas given off. When he put a candle into the gas, it burned extremely brightly. Perhaps Priestley's new air helped other substances burn. Could it explain why substances that burn gain weight?

To answer his questions, Lavoisier designed a new piece of equipment. It was an airtight reaction chamber and oven. He put liquid mercury (Hg) into the reaction chamber, put the chamber in the oven, and heated it for 12 days (1).

Three interesting things happened (2). The mercury turned into a reddish solid. The volume of air in the container decreased from 820 to 690 cubic centimeters (cm^3). And the new solid substance weighed more than the mercury!

Lavoisier concluded that the missing 130 cm^3 of air had combined with the shiny liquid mercury. The product was dull-red mercuric oxide. The additional mass of the mercuric oxide was exactly equal to the mass of the "missing" 130 cm^3 of air.

He then heated the solid mercuric oxide to a higher temperature. It turned back into liquid mercury and gave off the exact amount of gas that had been lost before.

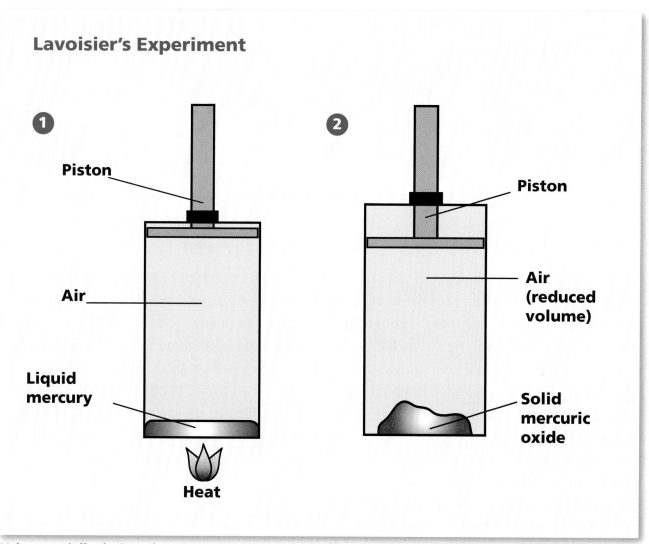

Lavoisier's Experiment

Using specially designed equipment, Lavoisier established the key principle of conservation of mass: the total mass of the products of a reaction is equal to the total mass of the reactants.

Investigation 9: *Reaction*

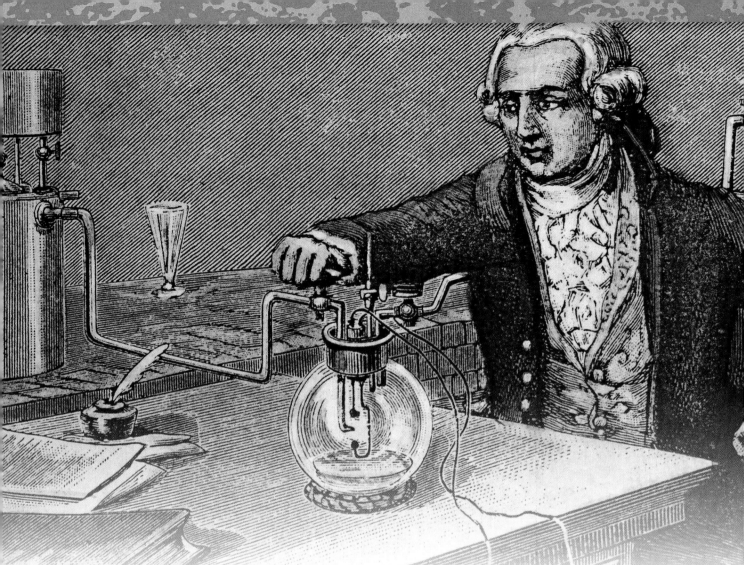

Lavoisier named oxygen (and many other elements) and discovered its vital role in combustion.

Oxygen Discovered

This change proved to Lavoisier that combustion does not burn phlogiston out of a substance. Instead, it combines the substance with part of the surrounding air. In 1785, Lavoisier said that his ideas should replace the idea of phlogiston. He later named the gas that causes combustion oxygen.

Some important chemists, including Priestley, disagreed with Lavoisier's conclusion. They still supported the phlogiston theory. But in the end, Lavoisier's careful experiments convinced a new generation of chemists that oxygen explained combustion.

Other Major Contributions

What impressed Lavoisier's supporters more than anything else was his careful attention to measurement. He spent a lot of his own money designing and building instruments. He developed balances as precise as many of those used today. With these instruments, he established the law of **conservation of mass**. That law states that matter is neither gained nor lost during a chemical reaction.

Lavoisier was also a showman. His equipment was flashy as well as precise. One piece he developed measured gas volume. It stood about 2 meters (m) tall and was made of gleaming brass. It was one of the most expensive pieces of chemistry apparatus anyone had ever seen. It cost more than $250,000 in today's dollars. Although its accuracy was no better than anyone else's laboratory equipment, its sheer expense and beauty made people begin to think of chemistry as serious science.

Many people still thought of chemistry as magic. But Lavoisier was changing that. In addition to his experiments, he helped develop the system of **chemical names** that still exists. This new vocabulary meant that chemists could communicate effectively. It helped make chemistry an independent, respected science.

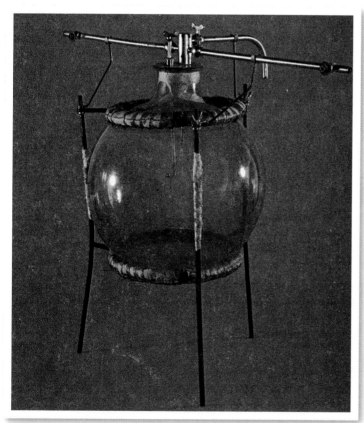

Lavoisier's gasometer was designed to measure the volume of gases. He admitted he relied on "very expensive and complicated instruments" for precise analysis and measurement.

Lavoisier listed the 33 known "elements." Though some are actually compounds or other things, like heat and light, he showed the need to name, classify, and organize the chemicals being discovered.

Investigation 9: Reaction **139**

To top it all off, in 1789, Lavoisier wrote the first textbook of chemistry, *The Elements of Chemistry*. He introduced the idea of elements, substances that could not be broken down into simpler substances. Lavoisier was wrong about some major ideas, however. He dismissed atoms as impossible. And he considered heat and light to be substances, rather than energy.

Did You Know?

The French Revolution broke out in 1789, the same year Lavoisier published *The Elements of Chemistry*. Because Lavoisier was a tax collector, he was arrested for crimes against the people of France. He was beheaded for this in 1794.

Think Questions

1. Why did mercuric oxide in Lavoisier's reaction chamber weigh more than the mercury metal?
2. Why was there less air in Lavoisier's reaction chamber after he heated the mercury for 12 days?
3. Why is Lavoisier considered to be the father of modern chemistry?

Lavoisier was arrested by leaders of the French Revolution because of his political involvements, mainly his role as a tax collector.

Organic Compounds

The year is 2050. You have been chosen to travel on the first human mission to Mars. Your mission is to search for life on another planet.

Finding new life might be extremely different, even unrecognizable. What evidence could you look for? Scientists do not yet know what life in other parts of the universe might look like. But they do know that all life on Earth relies on a category of substances called **organic compounds**. Organic compounds are made by living organisms, and they always contain carbon.

Carbon is the most important element in organic chemistry. Carbon atoms are unique in how they bond with each other to make long chains and rings. The many ways carbon atoms can combine results in an endless variety of molecules. No other element can bond in this same way. Those millions of different carbon-based organic molecules create the diversity of life on Earth.

All life that we know of is carbon based. So the search for life on Mars, or anywhere else in the universe, is a search for the element carbon.

The importance of carbon chaining cannot be overstated. Many scientists think that if there is life beyond Earth, it will be carbon based. No other atom has the ability to produce the complexity needed for life.

Getting Your Carbon

Did you have your carbon today? You need a constant supply to stay healthy and fit. But don't worry, you are getting plenty of carbon without even having to think about it. Here's how.

You have probably heard about how photosynthesis works. Plants and algae use water, light from the Sun, and carbon dioxide to make food. The food that plants make is sugar. Each sugar molecule is made of a carbon chain with hydrogen and oxygen atoms bonded to it. The carbon comes to the plant one atom at a time, in carbon dioxide gas. A series of complex reactions takes the carbon dioxide molecules apart and reassembles them into sugar.

The plant uses sugar for energy. The end products of that process are energy, water, and carbon dioxide. Some of the sugar produced by photosynthesis is also turned into other organic substances. Those substances include fats, proteins, and starches. These substances make all the other plant structures, like stems, leaves, flowers, and seeds.

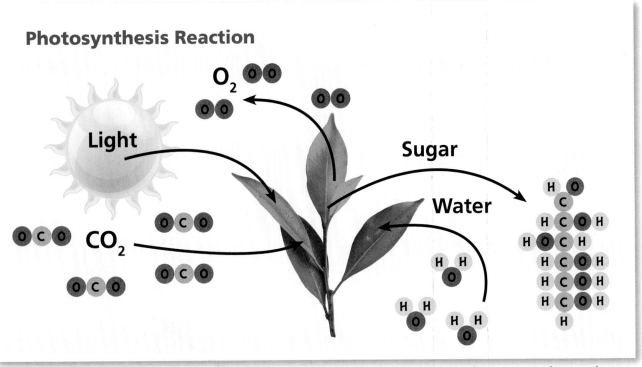

Water and carbon dioxide are the reactants in photosynthesis. Oxygen and sugar are the products.

Human beings don't work like plants, however. No matter how long you sit in the sunshine, you cannot make sugar out of carbon dioxide. So how do humans get carbon? We eat carbon-rich organic material. That means plants, plant parts, and animals. Your body breaks down the carbon-rich substances during digestion. It uses them to make the molecules you need to survive. You are a very advanced chemical factory, producing thousands of different carbon-based substances all the time.

Some of the carbon-based molecules are used for energy. A complex series of reactions takes apart the carbon chains and rearranges them into new substances. That process produces water and carbon dioxide. Water can be recycled, but carbon dioxide is a toxic waste. How do you get rid of that? Red blood cells carry the carbon dioxide molecules one by one to the lungs. There the carbon dioxide molecules are exhaled. And you have already proved that the carbon dioxide is returning to the atmosphere. Just blow some exhaled air through limewater, and watch the precipitate form.

The carbon-based molecules your cells are made of come from your food. So it's true: you are what you eat!

Investigation 9: Reaction

Hydrocarbons

Some organic compounds are made from only two elements, hydrogen and carbon. These substances are called **hydrocarbons**. They include natural gas, petroleum, and some coal. Combustion of these hydrocarbon fuels provides energy that we use in many ways.

Can you guess how some of these fuels formed? Under the right circumstances, dead organisms did not decay. Instead, they chemically changed into petroleum or coal over a long period of time.

One of the simplest hydrocarbons is methane (CH_4). Methane is a gas under normal atmospheric conditions. It is the main substance in natural gas, which is used in homes and businesses for cooking and heating. Propane (C_3H_8) is a hydrocarbon with a slightly larger molecule, made from 11 atoms. Propane is also gas under normal atmospheric conditions.

Propane under pressure changes into liquid. That property makes propane easy to store and transport. Easy transport makes propane a good fuel for people who live in remote areas, or who travel in recreational vehicles. It is also commonly used as fuel for gas grills. The fuel is kept in pressurized bottles or tanks. When the valve on the storage tank is opened, the propane turns into gas. The gas is used to cook dinner.

This table shows the chemical formulas and models of several hydrocarbons.

Simple Hydrocarbons

Substance name	Number of carbon atoms	Chemical formula
Methane	1	CH_4
Ethane	2	C_2H_6
Propane	3	C_3H_8
Butane	4	C_4H_{10}
Pentane	5	C_5H_{12}
Hexane	6	C_6H_{14}
Heptane	7	C_7H_{16}
Octane	8	C_8H_{18}

Propane fuel can be compressed from a gas to liquid and stored in a small space. So the tank beside this backyard "gas" grill contains liquid, not gas, propane.

High octane gasoline increases an engine's efficiency and performance. Higher octane gasoline is more expensive for this reason.

Octane

When a car pulls into a gas station, the driver has a choice. Which grade of gas is the best for the car? Regular is the lowest grade. If you check the pump, you may see that it is labeled 87 **octane**. What is octane? It is a substance, with the chemical formula C_8H_{18}. Octane is the main ingredient in gasoline. Gasoline that is rated 87 octane is 87 percent octane.

More expensive grades have higher ratings, maybe as high as 92 or 93 octane. Motorists who choose economical cars reach for the regular. Those with high-performance cars must use the high-octane fuels.

Under normal conditions, octane is a liquid. If it is ignited with a match, it will react with the oxygen in air and burn with a yellow flame. But if it is sprayed into a fine mist and ignited with a spark, it will explode. An explosion is a reaction that occurs extremely rapidly. It gives off a lot of light and thermal energy. Explosions inside the cylinders of an engine enable a car to move.

The octane reaction produces two products: carbon dioxide and water. Here is the equation. Is it balanced?

$$2C_8H_{18} + 25O_2 \rightarrow 16CO_2 + 18H_2O$$

Look at the atom representations of this equation. You can see the large number of particles involved in this important reaction.

Octane Reaction

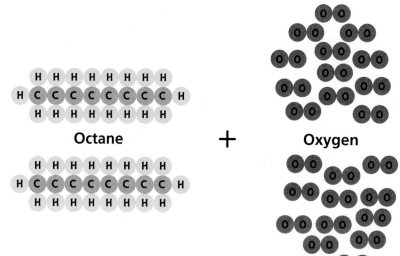
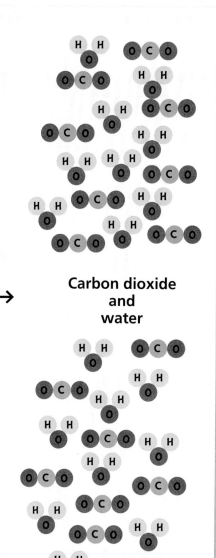

Octane + Oxygen → Carbon dioxide and water

The name octane is a clue to the molecular structure of octane: oct- means 8, indicating the number of carbon atoms present. In an octane reaction, 2 octane molecules combine with 25 oxygen molecules. What molecules are formed?

Did You Know?

Carbon dioxide is a greenhouse gas. As you can see in the octane reaction diagram, when cars burn octane-based fuel, they release carbon dioxide into the atmosphere. This causes atmospheric carbon dioxide levels to increase, which raises global average temperatures, causing climate change. Reducing fuel use is one way that people can decrease atmospheric carbon dioxide levels.

Think Questions

1. Why do scientists think that life on another planet might be carbon based?
2. What elements are in hydrocarbon compounds?

Careers in Chemistry

In middle school, it's hard to know what career you might choose. But there is a good chance that it will involve chemistry in some way.

As a pharmacologist, you might look for new drugs by studying how chemicals affect the human body. Forensic scientists collect materials at a crime scene. Chefs needs to understand the interactions of ingredients to come up with new recipes. They use chemistry when they solve problems or to create new things.

A biochemist might analyze tissue samples in a biotech lab to test interactions with substances and develop antibiotics. This is just one of many chemistry careers.

Some careers focus more directly on chemistry. Physical chemists investigate the structure of compounds and what happens in chemical reactions. Materials scientists try to improve substances. They might make lighter, stronger materials for cars and airplanes. Chemical engineers solve problems with producing and using substances. Their solutions might help purify drinking water, treat waste, and process food. Environmental chemists study how chemicals and technologies affect living things and their environment. Biochemists explore the molecules that support life, such as DNA and other important proteins. And organic chemists focus on the carbon-based substances in living things.

Donna Nelson, Organic Chemist

As an organic chemist, Donna Nelson (1954–) studies the rate of chemical reactions. She modifies a molecule to see how it affects the speed of a reaction. This information helps her discover properties of a substance and develop new substances.

Investigation 10: Limiting Factors

Many substances are ingredients in things that we use every day, from medicine to laundry soap. Each substance needs to be tested under different conditions. How does it react when combined with new substances?

The people who develop new products such as soap and medicine are called chemical engineers. As an organic chemist, Nelson provides these engineers with important information about their ingredients. Thus, her work helps make sure that all the ingredients of a new product work well together. Nelson says, "The work that I do is very exciting, and some days it feels like a big brain teaser. As a kid, I always loved brain teasers."

Dr. Donna Nelson's research focus is nanoscience, the study of matter and change on an extremely small scale. She also does work to support science education in American classrooms.

Products like soap have different properties based on the chemical ingredients selected.

Nelson grew up in a small town in Oklahoma. Her father was the town doctor. Her parents encouraged her curiosity about math and science. It was uncommon in her town for girls to go to a big university. And at the time, the University of Oklahoma had only a few Native American students. She says, "Going to a big university, where there were barely any Native Americans, was a big shock. It took time to adjust to being one of very few women and the only Native American in my department. I survived by being quiet in school, but very persistent."

Nelson is not quiet anymore! She has pushed for the media to portray science accurately. For example, she gave advice on scientific matters for a popular TV series. She has also worked hard to point out the status of women and minorities in the sciences at US universities. In 2015, she was elected president of the American Chemical Society.

Gertrude B. Elion, Biochemist

Have you heard of the disease leukemia? It is a blood cancer. Before 1950, half the children who got leukemia died in a few months. That was before Gertrude B. Elion (1918–1999) created a molecule that could attack the cancer.

Elion decided to become a scientist soon after high school graduation. After watching helplessly as her grandfather died of cancer, she resolved to find a cure for the disease. In 1937, Elion graduated from Hunter College in New York with a degree in chemistry. She applied for graduate studies at many colleges. Over and over, she heard that there were no jobs for women. At that time, women did not work in laboratories.

Elion and fellow scientist George Hitchings pioneered a new method of drug discovery. They designed new molecules with specific structures to block the growth of unwanted cells.

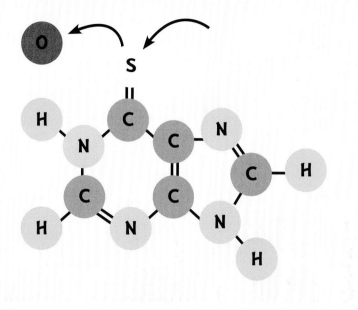

Elion's Replacement Reaction

Elion made the cancer-fighting 6-MP molecule by using a specific reaction to replace an oxygen atom with a sulfur atom.

In 1944, the United States was engaged in World War II. Many men were going off to war, and jobs were opening up for women. That year, Elion got her first research position with a drug company.

Elion referred to 1950 as her "Wow!" year. That year, she developed a molecule called 6-mercaptopurine (6-MP). Elion made the molecule with a reaction that replaced an oxygen atom with a sulfur atom. The 6-MP molecule helps boost the human immune system. It helps the body fight cancers like leukemia. Today, chemicals like 6-MP have increased the survival rate for children with leukemia to almost 90 percent.

Elion spent the next 30 years developing compounds to help the human body fight diseases. But she was interested in other things, too. She carefully studied the interactions between diseases and cells. She wanted to know why cancer cells were so successful. What were they doing, and what allowed them to do it? First, she figured out what the cancer needed to grow. Then, she used that information to produce a molecule that slowed or stopped the cancer.

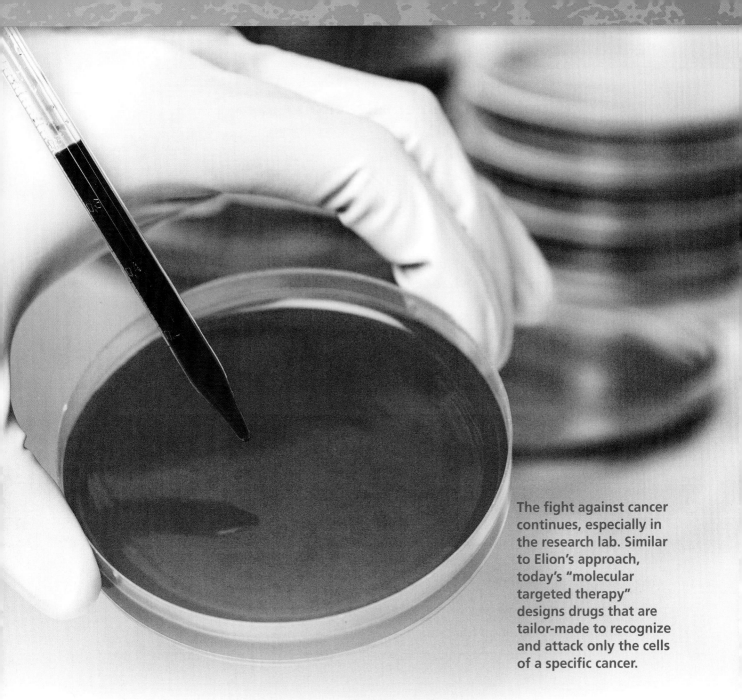

The fight against cancer continues, especially in the research lab. Similar to Elion's approach, today's "molecular targeted therapy" designs drugs that are tailor-made to recognize and attack only the cells of a specific cancer.

In 1988, Elion received the Nobel Prize in Physiology or Medicine for her many discoveries. This achievement was remarkable, because she had neither a medical degree nor a doctorate in chemistry. She won the prize through hard work, determination, and constant curiosity.

Elion was successful because she refused to lose sight of her dreams. "Don't be afraid of hard work," she once said in a lecture. "Nothing worthwhile comes easily. Don't let others discourage you or tell you that you can't do it. In my day, I was told women didn't go into chemistry. I saw no reason why we couldn't."

Think Questions

1. What careers interest you?
2. How do you think those careers might involve chemistry?

Element Hunters

How do you find a new element? That is a very interesting question.

In 1907, Ernest Rutherford (1871–1937) discovered the structure of the gold atom. His experiments suggested that atoms have a solid, positively charged **nucleus** with negatively charged **electrons** orbiting it. That was a breakthrough.

Using Rutherford's ideas, scientists figured out the structure of all atoms. With models for structures of the atoms, scientists could predict how atoms might combine to form molecules. They predicted that atoms in extended structures would line up in **well-ordered arrays**. Well-ordered arrays are repeating patterns, like marbles that all lay flat and line up in a box or oranges carefully stacked at the market.

A scanning tunneling microscope, with magnification up to 500,000X, is an instrument that shows 3-D images of the surface of a sample at the atomic level.

Investigation 10: Limiting Factors

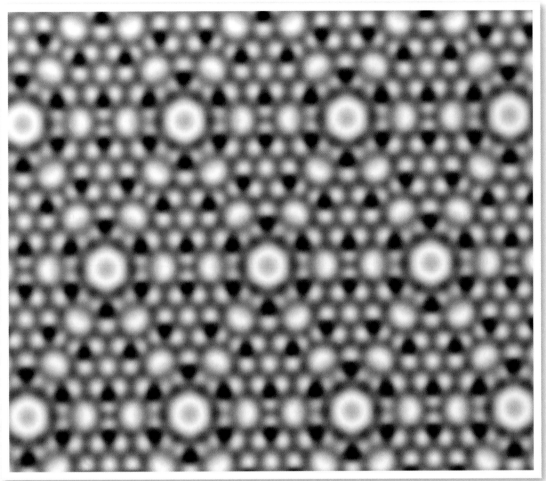

Each white dot in the image of silicon is one atom. The silicon atoms form an interesting pattern of circles.

For years, it was impossible to know whether the scientists' predictions were right. Atoms were way too small to see, even with the world's most powerful microscopes. This changed in 1981. A new kind of microscope was invented. The **scanning tunneling microscope** can create images of individual atoms. For the first time, scientists could see that their predictions were correct. Atoms *do* line up in orderly arrays.

The Search for New Elements

By 1940, all 90 naturally occurring elements had been discovered and described. The periodic table went from element 1, hydrogen, to element 92, uranium. Elements 43 and 61 had not been discovered. They just did not seem to exist on Earth. And no elements existed beyond element 92, uranium. Element number is also called **atomic number**.

Scientists knew that if they could put one more **proton** in the nucleus of a uranium atom, the result would be a new element, element 93. But how could they do that? In 1931, a young professor at the University of California at Berkeley, Ernest O. Lawrence (1901–1958), invented a device called the **cyclotron**. The cyclotron made it possible to create new elements. It used electricity to speed up protons and other atomic particles. The particles could move so fast that they would crash into the nucleus of uranium and stick there. In 1940, the cyclotron was used to make element 93, neptunium.

Between 1940 and 1974, elements 93 through 106 were created in the labs at Berkeley. Two scientists led the research teams. They were Glenn Seaborg (1912–1999) and Albert Ghiorso (1915–2010). They decided which atoms to collide in order to make new elements. They also designed bigger and bigger cyclotrons. By the 1970s, a single cyclotron would fill one huge building.

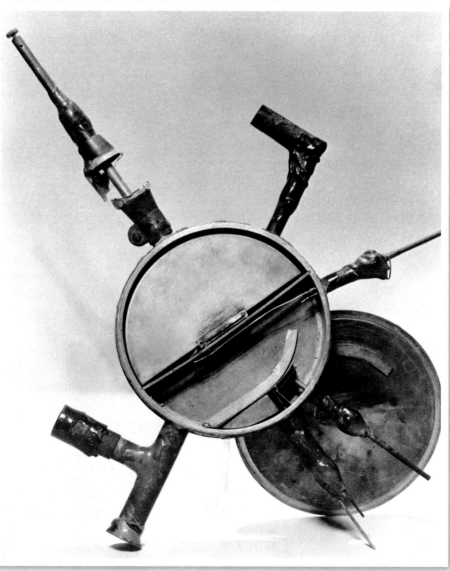

Lawrence's first cyclotron, only 28 cm in diameter, used high voltage to accelerate particles in a spiral path. He had discovered a way to "smash" atoms together to create new elements.

Investigation 10: Limiting Factors

In the 1980s and 1990s, elements 107 through 112 were discovered. Throughout the first part of the 2000s elements 113 through 118 were created. But even though scientists have been trying to create element 119 and 120, they have not had success. Will they be able to do it? Pay attention to the science news to find out.

Of the current 118 discovered elements, 90 occur naturally on Earth. Only 80 elements are stable. The others are radioactive. They change chemically into lighter elements over time. The newly discovered elements decay in just milliseconds. This makes it hard to prove they ever existed.

> **Take Note**
>
> **Do you think there is a limit to the size of atoms scientists can make?**

James A. Harris

In 1969, James A. Harris (1932–2000) was on the team of scientists who discovered elements 104 and 105. He was the first African American scientist to discover new elements.

Harris grew up in Waco, Texas, and Oakland, California. He earned his college degree in chemistry, but he faced racial discrimination while applying for jobs as a scientist in the 1950s. He later talked about how interviewers did not believe he was qualified when they first met him. He didn't give up, and found a job in a research lab in California. Five years later, he began working at the national laboratory in Berkeley, California.

By 1934, Lawrence's cyclotron had tripled in size. The use of the cyclotron to bombard atoms has made possible the discovery, or creation, of 26 new elements.

James Harris, a nuclear chemist at Berkeley's Lawrence Radiation Lab, co-discovered the two elements now named Rutherfordium (104) and Dubnium (105).

At the national lab in Berkeley, Harris was a head of production for materials. These materials would be used in the accelerator to create new elements. He excelled at the challenging work, and was known among his colleagues for creating extremely high-quality materials.

Harris also designed the target in the experimental system. The target is necessary to identify the new elements produced by collisions.

Harris had a passion for his career. He wanted to encourage and support other young scientists. He visited schools and universities to meet with science and engineering students. He encouraged students from all backgrounds, including other African American youth, to pursue their dreams despite the challenges they may face. He earned awards for this outreach work from the Urban League and the National Organization for the Professional Advancement of Black Chemists and Chemical Engineers.

Investigation 10: Limiting Factors

Ernest O. Lawrence

The discovery of 26 elements beyond uranium has been possible because of the cyclotron invented by Lawrence. His tiny cyclotron sparked a new branch of science. Lawrence accomplished a lot as a scientist. He received the Nobel Prize in 1939, the first time it was awarded to someone at the University of California. Two national research laboratories are named for him.

After his death, Lawrence's family and fellow scientists wanted to build a memorial in his honor. They decided to build a science museum, the Lawrence Hall of Science. Every year, thousands of families and school classes visit the museum. The Lawrence Hall of Science is an active, ever-changing science experience. It is a fitting memorial to a man who started a science revolution.

Think Questions

1. **How do scientists know that atoms and molecules line up in well-ordered arrays?**
2. **Explain why the periodic table has over 110 elements when only 90 occur naturally on Earth.**

Lawrence Hall of Science, dedicated in 1968, is the only US science center that is part of a public research university. The stated goal is to inspire and foster science learning for all.

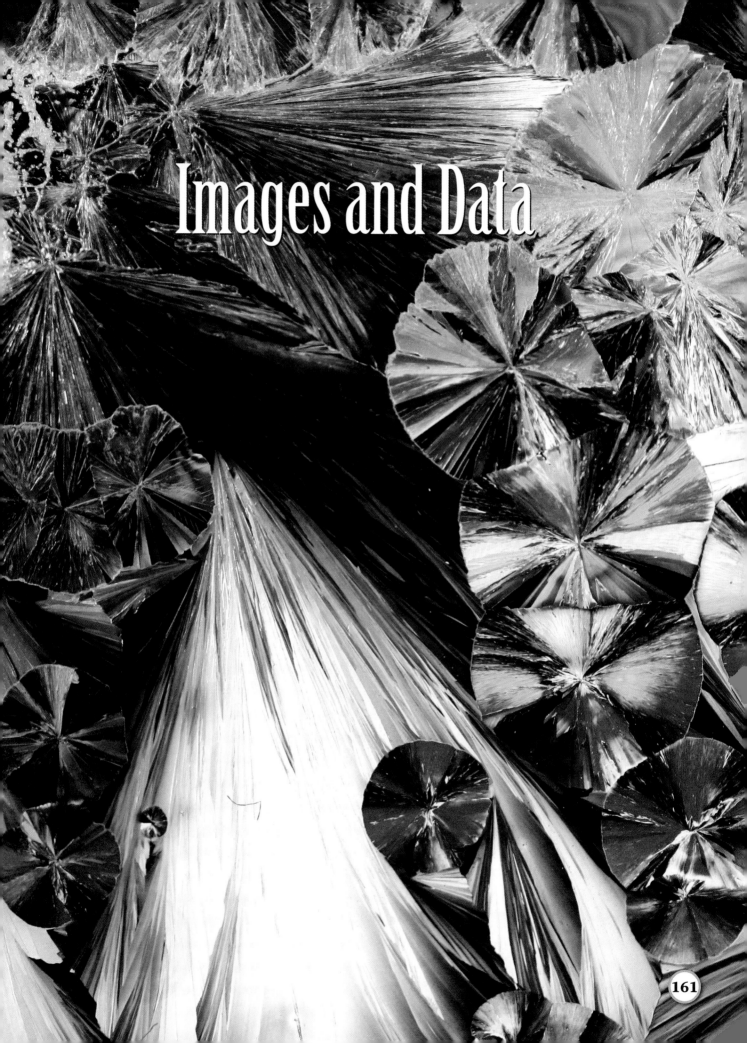
Images and Data

Images and Data Table of Contents

Investigation 1: Substances
White Substances Information **165**

Investigation 2: Elements
The Periodic Table of the Elements **174**
The Elements in Alphabetical Order **176**
The Most Common Elements **178**

Investigation 9: Reaction
Atoms and Compounds. **180**
Compound Structure **181**

References
Science Practices. **182**
Engineering Practices. **183**
Engineering Design Process. **184**
Science Safety Rules **185**
Glossary **186**
Index **190**

White Substances Information

Ascorbic Acid

Ascorbic acid ($C_6H_8O_6$) is also known as vitamin C. It is an essential part of the human diet. All vertebrates (animals with backbones), except primates (including humans) and guinea pigs, make their own vitamin C. Guinea pigs and primates must get it from foods. Good sources are citrus fruit, tomatoes, and liver.

Vitamin C helps body tissues grow and heal. It helps make an important protein called collagen. Collagen is found in bones, cartilage, soft tissues, and teeth. Without vitamin C, joints hurt and grow weak, and gums bleed. Teeth loosen, making it very difficult and painful to eat. The final result is death.

The connection between these symptoms, called scurvy, and vitamin C was unknown until the late 1700s. A British naval doctor observed that eating citrus fruits cured scurvy. From that time on, sailors were required to drink lime or lemon juice to prevent scurvy.

You may have heard that a massive dose of vitamin C can prevent the common cold. According to medical studies, however, this is not true. The health benefits associated with vitamin C do not include cold prevention.

Ascorbic acid, or vitamin C, is essential for the healthy development and function of many parts of the body. Getting it from a diet rich in fruits and vegetables is best, but granulated ascorbic acid can be dissolved and taken as a supplement.

Calcium Carbonate

Calcium carbonate ($CaCO_3$) is a common mineral. It is found in sedimentary rocks, such as chalk, limestone, and marble. Carbonate rocks account for about 4 percent of the mass of Earth's crust. Calcium carbonate is also important in ocean ecosystems. Snails, oysters, and clams make their shells out of it. When you see a coral reef, you are looking at the calcium carbonate skeletons of millions of tiny coral animals.

Calcium carbonate is an inexpensive source of calcium. It is used to make the calcium pills that people take to strengthen bones. It is also used in antacids to relieve acid indigestion after a big meal. Calcium carbonate neutralizes excess hydrochloric acid in our stomachs.

And don't forget the chalkboard. Chalk is used in classrooms around the world. Writing chalk is usually made of calcium carbonate.

Calcium carbonate is found around the world in sedimentary rocks. It is widely used in the construction industry, both as a building material and in making cement.

Calcium Chloride

Calcium chloride ($CaCl_2$) is a salt. It is not the same salt you use to flavor foods. Calcium chloride has two useful properties. When it dissolves, it releases thermal energy. This makes calcium chloride good for melting ice on roads and sidewalks. As the salt dissolves in the melted ice, it lowers the freezing point of water. The solution does not freeze as quickly as pure water. Calcium chloride has another benefit as road salt. It damages plants growing at the sides of roads much less than other salts do.

Calcium chloride is hygroscopic. That means it absorbs water. Because it soaks up water so efficiently, it is used to dry air and other gases. It is also spread on dirt roads. The calcium chloride absorbs water, which holds down the dust.

Calcium chloride is added to concrete to make it set faster and cure harder. It is used as a food additive (in pickles particularly) to add a salty taste without increasing the sodium content of the food. This is important for people who are on low-sodium diets and cannot eat table salt, which is sodium chloride.

Used in many industries, calcium chloride is best known for its "road work." As a de-icer, it helps keep winter pavements bare and safe.

Citric Acid

Citric acid ($C_6H_8O_7$) is found in almost all plants and in many animal tissues and fluids. It is important in animal metabolism. One good source of citric acid is citrus fruits, like lemons, oranges, tangerines, and grapefruits. Citric acid is also found in strawberries, apples, peaches, rice, soybeans, and wheat.

Much of the citric acid used in the United States finds its way into food. It is used to preserve foods, like jams and jellies. It is also used to give texture to processed cheese. Citric acid, like all acids, has a sharp, sour taste. It is added to candies and soft drinks to give them a sour zing. Citric acid is sometimes called "sour salt." When you pop a sour candy in your mouth and feel your cheeks pucker, you are having a citric acid moment.

Citric acid occurs naturally, and most abundantly, in citrus fruits. It appears on nutrition labels in almost every grocery aisle as an additive that preserves, blends, and flavors food—from ice cream and cheeses to soups and soft drinks.

Magnesium Sulfate

Ahhh, soak those sore feet in a nice warm bath of Epsom salts. For centuries, people have soothed their sore feet in magnesium sulfate ($MgSO_4$) mineral springs.

Perhaps the most famous magnesium sulfate spring is in Epsom, England, hence the name Epsom salts. In the early 1600s, a farmer noticed that his thirsty cattle would not drink at a certain spring. The water was very bitter. But the magnesium sulfate salts were found to be relaxing and medicinal.

Today, magnesium sulfate is sold in drugstores as a soaking agent for bruised, tired feet. It is also relieves constipation. It is used in fertilizers as a source of magnesium, and in detergents and soaps. It is even added to purified water to give it some taste.

Magnesium sulfate is a bitter-tasting mineral compound. Taken orally or absorbed through the skin by soaking, magnesium sulfate has several health benefits.

Sodium Bicarbonate

Did you ever see a model volcano erupt? The "lava" was probably produced by mixing sodium bicarbonate ($NaHCO_3$) and vinegar. Sodium bicarbonate's common name is baking soda. It makes biscuits and other baked goods light and fluffy. When sodium bicarbonate reacts with acid, carbon dioxide gas forms. That gas makes the foam in the model volcano and causes biscuit dough to rise.

Sodium bicarbonate is an ingredient in many brands of toothpaste. Why is it helpful? The bacteria that grow in your mouth give off acid as a waste product. That acid eats away at the outer layer of your teeth and causes them to decay. A toothpaste with sodium bicarbonate neutralizes the acid. In fact, sodium bicarbonate is so good at fighting tooth decay that some people use it alone to brush their teeth.

Sodium bicarbonate is also good for indigestion caused by excess stomach acid. It is an ingredient in many over-the-counter antacids. Antacid literally means "against acid." Sodium bicarbonate reacts with that extra acid, forming neutral products: carbon dioxide gas, table salt, and water.

Sodium bicarbonate is a chemical compound that usually appears as the fine white powder, baking soda. Its acid-neutralizing properties are useful on teeth and in tummies.

Sodium Carbonate

Sodium carbonate (Na_2CO_3) is also known as washing soda or soda ash. It is found naturally as trona ore in a few parts of the world. The largest deposit of trona is in Wyoming.

In the 1700s, sodium carbonate, recovered from seaweed ashes, was used to make glass and soap. Today, a chemical process can be used to make synthetic sodium carbonate. However, this process produces a number of hazardous wastes. The preferred method of obtaining sodium carbonate is still mining and refining it from natural ore.

Sodium carbonate is important in glassmaking. To make glass, you need to melt sand (silicon dioxide). Sand melts at 1,700°C. If you add sodium carbonate to the sand, it lowers the melting temperature, making it less expensive to produce glass.

Glass production is the largest use for sodium carbonate, but sodium carbonate has many other uses. It is used in manufacturing detergents and soaps, making paper, and treating waste water.

Unlike sodium bicarbonate, sodium carbonate is toxic and may be harmful if swallowed. Around the house it is found in cleaning products; it helps soaps lather and do their jobs.

Sodium Chloride

Sodium chloride (NaCl), the salt you put on food, is so important that it has played a role in the development of civilization. Salt has led to war and has served as money. It is considered a universal symbol of hospitality. Why is salt so important? Our bodies cannot make it but must have it. Without sodium and chlorine, our bodies cannot function properly.

Sodium is essential for muscle movement, heartbeat regulation, and nerve function. Chlorine is in stomach acid, making digestion possible. It also controls the movement of water into and out of cells.

Salt is an excellent preservative, used to keep meat, fish, and vegetables from spoiling. Food that has been salted and cured can remain edible for months. Pickling foods, like cucumbers (pickles), is another method of food preservation. Food is first soaked in brine (salt water), followed by vinegar. Before refrigerators, salt was the most important preservative.

Today, sodium chloride is used for much more than preserving foods. The salt industry claims that salt has more than 14,000 uses! Most salt used in the United States goes to make other substances, like chlorine, sodium carbonate, and hydrochloric acid.

Where we do get salt? The oldest method of salt production is still in use: solar evaporation of salt water. Salt is also mined from underground deposits.

Sucrose

Sucrose ($C_{12}H_{22}O_{11}$) is best known as the white sugar used to sweeten foods. It comes from plants. The two most important source plants are sugarcane and sugar beets.

About 70 percent of the sugar produced in the world is extracted from sugarcane, a very tall grass that looks like bamboo. It is grown in tropical regions around the world. The rest comes from sugar beets. Sugar beets are roots that are grown in northern, cooler climates. They look like fat, white carrots.

Why is sugar so important? It tastes good and is a source of energy in the human diet. There are 16 food calories in every teaspoon of sucrose. If you are like most Americans, you eat 20 kilograms (kg) of sugar per year. That is ten 2 kg bags, or about 57 grams of sugar per day. That means that you eat almost 200 food calories per day of pure sucrose. Your body breaks down the sucrose into glucose, which cells use as their most important fuel source. Sugar, in moderation, is an important part of our diet.

Sugar also has other uses. It is used in large quantities as a preservative in jams and jellies. It is also a food source for yeast in bread dough.

Table sugar is crystallized sucrose. The body cannot use this carbohydrate for energy, though, until it is broken down during digestion into simpler sugars, like glucose.

Investigation 1: Substances

The Periodic Table of the Elements

1 H Hydrogen								
3 Li Lithium	4 Be Beryllium							
11 Na Sodium	12 Mg Magnesium							
19 K Potassium	20 Ca Calcium	21 Sc Scandium	22 Ti Titanium	23 V Vanadium	24 Cr Chromium	25 Mn Manganese	26 Fe Iron	27 Co Cobalt
37 Rb Rubidium	38 Sr Strontium	39 Y Yttrium	40 Zr Zirconium	41 Nb Niobium	42 Mo Molybdenum	43 Tc Technetium	44 Ru Ruthenium	45 Rh Rhodium
55 Cs Cesium	56 Ba Barium	71 Lu Lutetium	72 Hf Hafnium	73 Ta Tantalum	74 W Tungsten	75 Re Rhenium	76 Os Osmium	77 Ir Iridium
87 Fr Francium	88 Ra Radium	103 Lr Lawrencium	104 Rf Rutherfordium	105 Db Dubnium	106 Sg Seaborgium	107 Bh Bohrium	108 Hs Hassium	109 Mt Meitnerium

57 La Lanthanum	58 Ce Cerium	59 Pr Praseodymium	60 Nd Neodymium	61 Pm Promethium	62 Sm Samarium
89 Ac Actinium	90 Th Thorium	91 Pa Protactinium	92 U Uranium	93 Np Neptunium	94 Pu Plutonium

Alkali Metals
Alkaline Earths
Metals
Nonmetals
Halides
Noble gases
Metalloids
Lanthanides
Actinides

								2 **He** Helium
			5 **B** Boron	6 **C** Carbon	7 **N** Nitrogen	8 **O** Oxygen	9 **F** Fluorine	10 **Ne** Neon
			13 **Al** Aluminum	14 **Si** Silicon	15 **P** Phosphorus	16 **S** Sulfur	17 **Cl** Chlorine	18 **Ar** Argon
28 **Ni** Nickel	29 **Cu** Copper	30 **Zn** Zinc	31 **Ga** Gallium	32 **Ge** Germanium	33 **As** Arsenic	34 **Se** Selenium	35 **Br** Bromine	36 **Kr** Krypton
46 **Pd** Palladium	47 **Ag** Silver	48 **Cd** Cadmium	49 **In** Indium	50 **Sn** Tin	51 **Sb** Antimony	52 **Te** Tellurium	53 **I** Iodine	54 **Xe** Xenon
78 **Pt** Platinum	79 **Au** Gold	80 **Hg** Mercury	81 **Tl** Thallium	82 **Pb** Lead	83 **Bi** Bismuth	84 **Po** Polonium	85 **At** Astatine	86 **Rn** Radon
110 **Ds** Darmstadtium	111 **Rg** Roentgenium	112 **Cn** Copernicium	113 **Nh** Nihonium	114 **Fl** Flerovium	115 **Mc** Moscovium	116 **Lv** Livermorium	117 **Ts** Tennessine	118 **Og** Oganesson

63 **Eu** Europium	64 **Gd** Gadolinium	65 **Tb** Terbium	66 **Dy** Dysprosium	67 **Ho** Holmium	68 **Er** Erbium	69 **Tm** Thulium	70 **Yb** Ytterbium	
95 **Am** Americium	96 **Cm** Curium	97 **Bk** Berkelium	98 **Cf** Californium	99 **Es** Einsteinium	100 **Fm** Fermium	101 **Md** Mendelevium	102 **No** Nobelium	

The Elements in Alphabetical Order

Element name	Symbol	Atomic number	Date of discovery
Actinium	Ac	89	1899/1902
Aluminum	Al	13	1825
Americium	Am	95	1944
Antimony	Sb	51	Prehistoric
Argon	Ar	18	1894
Arsenic	As	33	Middle Ages
Astatine	At	85	1940
Barium	Ba	56	1808
Berkelium	Bk	97	1949
Beryllium	Be	4	1797
Bismuth	Bi	83	1753
Bohrium	Bh	107	1981
Boron	B	5	1808
Bromine	Br	35	1826
Cadmium	Cd	48	1817
Calcium	Ca	20	1808
Californium	Cf	98	1950
Carbon	C	6	Prehistoric
Cerium	Ce	58	1803
Cesium	Cs	55	1860
Chlorine	Cl	17	1774
Chromium	Cr	24	1797
Cobalt	Co	27	1735
Copernicium	Cn	112	1996
Copper	Cu	29	Prehistoric
Curium	Cm	96	1944
Darmstadtium	Ds	110	1994
Dubnium	Db	105	1970
Dysprosium	Dy	66	1886
Einsteinium	Es	99	1952
Erbium	Er	68	1842
Europium	Eu	63	1901
Fermium	Fm	100	1952
Flerovium	Fl	114	1998
Fluorine	F	9	1886
Francium	Fr	87	1939
Gadolinium	Gd	64	1880
Gallium	Ga	31	1875
Germanium	Ge	32	1886
Gold	Au	79	Prehistoric
Hafnium	Hf	72	1923
Hassium	Hs	108	1984
Helium	He	2	1868
Holmium	Ho	67	1878
Hydrogen	H	1	1766
Indium	In	49	1863
Iodine	I	53	1811
Iridium	Ir	77	1803
Iron	Fe	26	Prehistoric
Krypton	Kr	36	1898
Lanthanum	La	57	1839
Lawrencium	Lr	103	1961
Lead	Pb	82	Prehistoric
Lithium	Li	3	1817
Livermorium	Lv	116	2000
Lutetium	Lu	71	1907
Magnesium	Mg	12	1808
Manganese	Mn	25	1774
Meitnerium	Mt	109	1982
Mendelevium	Md	101	1955
Mercury	Hg	80	Prehistoric
Molybdenum	Mo	42	1778

Element name	Symbol	Atomic number	Date of discovery
Moscovium	Mc	115	2010
Neodymium	Nd	60	1885
Neon	Ne	10	1898
Neptunium	Np	93	1940
Nickel	Ni	28	1751
Nihonium	Nh	113	2004
Niobium	Nb	41	1801
Nitrogen	N	7	1772
Nobelium	No	102	1958
Oganesson	Og	118	2002
Osmium	Os	76	1803
Oxygen	O	8	1774
Palladium	Pd	46	1803
Phosphorus	P	15	1669
Platinum	Pt	78	1735
Plutonium	Pu	94	1940
Polonium	Po	84	1898
Potassium	K	19	1807
Praseodymium	Pr	59	1885
Promethium	Pm	61	1945
Protactinium	Pa	91	1913
Radium	Ra	88	1898
Radon	Rn	86	1900
Rhenium	Re	75	1925
Rhodium	Rh	45	1803
Roentgenium	Rg	111	1994
Rubidium	Rb	37	1861
Ruthenium	Ru	44	1844
Rutherfordium	Rf	104	1969
Samarium	Sm	62	1879
Scandium	Sc	21	1878
Seaborgium	Sg	106	1974
Selenium	Se	34	1817
Silicon	Si	14	1824
Silver	Ag	47	Prehistoric
Sodium	Na	11	1807
Strontium	Sr	38	1808
Sulfur	S	16	Prehistoric
Tantalum	Ta	73	1802
Technetium	Tc	43	1937
Tellurium	Te	52	1782
Tennessine	Ts	117	
Terbium	Tb	65	1843
Thallium	Tl	81	1861
Thorium	Th	90	1828
Thulium	Tm	69	1879
Tin	Sn	50	Prehistoric
Titanium	Ti	22	1791
Tungsten	W	74	1783
Uranium	U	92	1841
Vanadium	V	23	1801
Xenon	Xe	54	1898
Ytterbium	Yb	70	1878
Yttrium	Y	39	1794
Zinc	Zn	30	Prehistoric
Zirconium	Zr	40	1789

The Most Common Elements

Universe

Element name	Symbol	Percent by mass
Hydrogen	H	74.99%
Helium	He	23.00%
Oxygen	O	1.00%
Carbon	C	0.50%
Neon	Ne	0.13%
Iron	Fe	0.11%
Nitrogen	N	0.10%
Silicon	Si	0.07%
Magnesium	Mg	0.06%
Sulfur	S	0.05%

Sun/star

Element name	Symbol	Percent by mass
Hydrogen	H	75.23%
Helium	He	23.07%
Oxygen	O	0.90%
Carbon	C	0.30%
Neon	Ne	0.10%
Iron	Fe	0.10%
Nitrogen	N	0.10%
Silicon	Si	0.09%
Magnesium	Mg	0.07%
Sulfur	S	0.04%

Earth's crust

Element name	Symbol	Percent by mass
Oxygen	O	45.50%
Silicon	Si	27.20%
Aluminum	Al	8.30%
Iron	Fe	6.20%
Calcium	Ca	4.70%
Magnesium	Mg	2.80%
Sodium	Na	2.00%
Potassium	K	1.30%
Carbon	C	0.20%
Hydrogen	H	0.10%

Ocean

Element name	Symbol	Percent by mass
Oxygen	O	85.83%
Hydrogen	H	10.80%
Chlorine	Cl	1.99%
Sodium	Na	1.11%
Magnesium	Mg	0.13%
Sulfur	S	0.09%
Potassium	K	0.04%
Bromine	Br	0.01%
Carbon	C	trace
Strontium	Sr	trace
Boron	B	trace

Human

Element name	Symbol	Percent by mass
Oxygen	O	65.00%
Carbon	C	18.00%
Hydrogen	H	10.80%
Nitrogen	N	3.00%
Calcium	Ca	1.50%
Phosphorous	P	1.00%
Potassium	K	0.20%
Sulfur	S	0.20%
Chlorine	Cl	0.20%
Sodium	Na	0.10%

Trace elements include magnesium, iron, cobalt, zinc, iodine, selenium, and fluorine.

Atmosphere

Element name	Symbol	Percent by mass
Permanent Gases		
Nitrogen	N_2	78.08%
Oxygen	O_2	20.95%
Argon	Ar	0.93%
Neon	Ne	trace
Helium	He	trace
Krypton	Kr	trace
Hydrogen	H_2	trace
Xenon	Xe	trace
Variable Gases		
Water vapor	H_2O	0.25%
Carbon dioxide	CO_2	0.04%
Ozone	O_3	0.01%

Atoms and Compounds

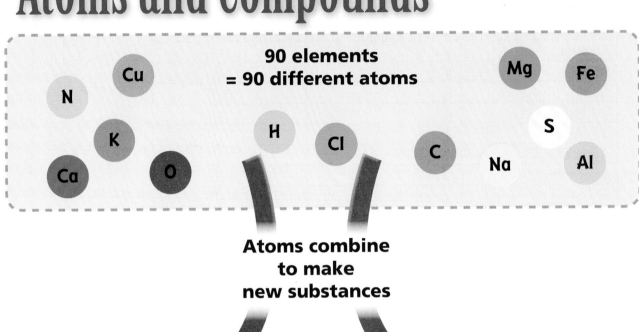
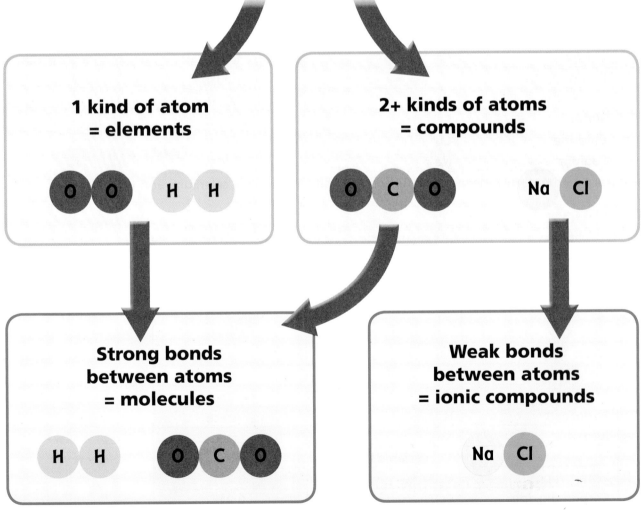

Compound Structure

Some compounds form molecules.

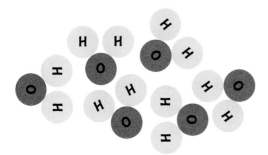

water (liquid)

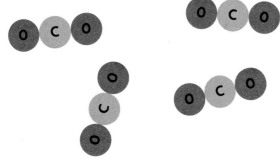

carbon dioxide (gas)

Some compounds form unique extended structures such as crystals.

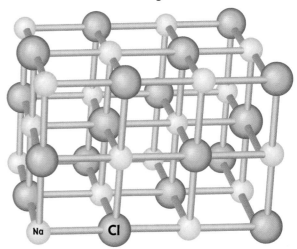

sodium chloride (solid)

What shape crystals does sodium chloride form?

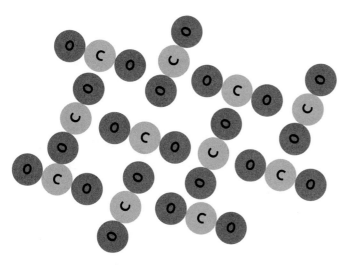

carbon dioxide (solid)

Investigation 9: Reaction 181

Science Practices

1. **Asking questions.** Scientists ask questions to guide their investigations. This helps them learn more about how the world works.
2. **Developing and using models.** Scientists develop models to represent how things work and to test their explanations.
3. **Planning and carrying out investigations.** Scientists plan and conduct investigations in the field and in laboratories. Their goal is to collect data that test their explanations.
4. **Analyzing and interpreting data.** Patterns and trends in data are not always obvious. Scientists make tables and graphs. They use statistical analysis to look for patterns.
5. **Using mathematics and computational thinking.** Scientists measure physical properties. They use computation and math to analyze data. They use mathematics to construct simulations, solve equations, and represent different variables.
6. **Constructing explanations.** Scientists construct explanations based on observations and data. An explanation becomes an accepted theory when there are many pieces of evidence to support it.
7. **Engaging in argument from evidence.** Scientists use argumentation to listen to, compare, and evaluate all possible explanations. Then they decide which best explains natural phenomena.
8. **Obtaining, evaluating, and communicating information.** Scientists must be able to communicate clearly. They must evaluate others' ideas. They must convince others to agree with their theories.

Are you a scientist?

Engineering Practices

1. **Defining problems.** Engineers ask questions to make sure they understand the problems they are trying to solve. They need to understand the constraints that are placed on their designs.

2. **Developing and using models.** Engineers develop and use models to represent systems they are designing. Then they test their models before building the actual object or structure.

3. **Planning and carrying out investigations.** Engineers plan and conduct investigations. They need to make sure that their designed systems are durable, effective, and efficient.

4. **Analyzing and interpreting data.** Engineers collect and analyze data when they test their designs. They compare different solutions. They use the data to make sure that they match the given criteria and constraints.

5. **Using mathematics and computational thinking.** Engineers measure physical properties. They use computation and math to analyze data. They use mathematics to construct simulations, solve equations, and represent different variables.

6. **Designing solutions.** Engineers find solutions. They propose solutions based on desired function, cost, safety, how good it looks, and meeting legal requirements.

7. **Engaging in argument from evidence.** Engineers use argumentation to listen to, compare, and evaluate all possible ideas and methods to solve a problem.

8. **Obtaining, evaluating, and communicating information.** Engineers must be able to communicate clearly. They must evaluate other's ideas. They must convince others of the merits of their designs.

Are you an engineer?

Engineering Design Process

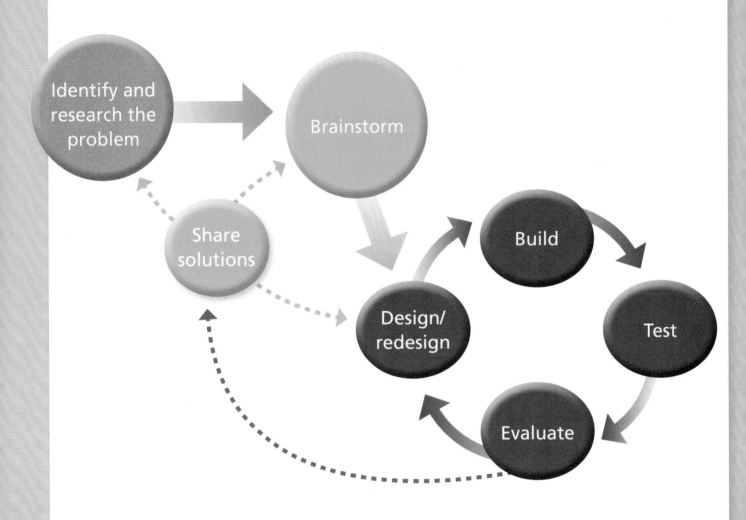

Science Safety Rules

1. Always follow the safety procedures outlined by your teacher. Follow directions, and ask questions if you're unsure of what to do.

2. Never put any material in your mouth. Do not taste any material or chemical unless your teacher specifically tells you to do so.

3. Do not smell any unknown material. If your teacher asks you to smell a material, wave a hand over it to bring the scent toward your nose.

4. Avoid touching your face, mouth, ears, eyes, or nose while working with chemicals, plants, or animals. Tell your teacher if you have any allergies.

5. Always wash your hands with soap and warm water immediately after using chemicals (including common chemicals, such as salt and dyes) and handling natural materials or organisms.

6. Do not mix unknown chemicals just to see what might happen.

7. Always wear safety goggles when working with liquids, chemicals, and sharp or pointed tools. Tell your teacher if you wear contact lenses.

8. Clean up spills immediately. Report all spills, accidents, and injuries to your teacher.

9. Treat animals with respect, caution, and consideration.

10. Never use the mirror of a microscope to reflect direct sunlight. The bright light can cause permanent eye damage.

Glossary

alchemy the prescientific investigation of substances, including the search for ways to change common metals into gold

atmosphere gases surrounding a planet

atom the smallest particle of an element

atomic number the number assigned to an element, based on the number of protons in the nucleus of its atom

biomagnification the process where a chemical is taken in by an organism due to consuming an organism that already has the chemical in their body

bond an attractive force between atoms

burning a chemical reaction in which a substance reacts with oxygen to form a new product

calorie the unit of energy that will raise the temperature of 1 gram of water 1 degree Celsius

carbohydrate a group of carbon-based nutrients, including sugars and starches

carbon dioxide a compound made of bonded carbon and oxygen atoms; CO_2

chemical equation a representation of a chemical reaction using chemical formulas

chemical formula a code that represents the number and kinds of atoms in one particle of a substance

chemical name the name chemists use for a substance, which refers to the names of the elements in that substance

chemical property a characteristic of a substance that determines how it interacts with other substances

chemical reaction a process during which the atoms of starting substances (reactants) rearrange to form new substances (products)

climate change the change in climate due to increased energy usage and greenhouse gases

combustion a chemical reaction, commonly called burning

common name the everyday-language name of a substance

compound a substance defined by a particle composed of two or more different kinds of atoms

compressed reduced in volume as a result of applied pressure

concentrated a solution with a large amount of solute dissolved in a small amount of solvent

concentration the amount of solute dissolved in a measure of solvent

condensation the change of phase from gas to liquid

conduction the transfer of energy (heat) from one particle to another as a result of contact

conservation of energy a scientific law stating that no energy is created or destroyed during energy transfers

conservation of mass a scientific law stating that no matter is created or destroyed during a reaction

conserved unchanged during a process

constraint a restriction or limitation

contraction the reduction of volume of a sample of matter as a result of cooling

cooling energy transfer that decreases the kinetic energy of a substance's particles

core the most dense, central layer of Earth, composed mostly of iron and nickel

criterion (plural: **criteria**) requirement

crude oil a material made of mostly carbon and hydrogen; also known as petroleum

crust Earth's hard outer layer of solid rock

crystal a structure formed by networks of atoms arranged in repeating patterns. Crystal shape is also a physical property that helps to identify a substance.

cyclotron an instrument used to create new elements

density the ratio of mass and volume in a sample of matter

deposit the change of phase from gas directly to solid

dilute a solution with a small amount of solute dissolved in a large amount of solvent

dissolve to mix one substance uniformly into another substance at the particle level

dry ice the solid phase of carbon dioxide

electron a subatomic particle with a negative charge

element a fundamental substance that cannot be broken into simpler substances by chemical or physical processes

energy transfer the movement of energy from one location to another

engineer someone who designs solutions based on scientific findings

engineering problem a challenge that engineers seek to solve by designing and testing solutions

equilibrium a condition in which a system is experiencing no net change

evaporation the change of phase from liquid to gas

exothermic reaction chemical reaction that transfers energy to the environment in the forms of light, thermal energy, and/or sound

expansion an increase of volume

extinct a group of organisms with no living members

force a push or a pull

freeze to change phase from liquid to solid

freezing point the temperature at which a liquid changes to a solid; different for each substance

fundamental simple and basic

gas a phase of matter that has no definite shape or volume. Particles of gas fly independently through space.

gaseous existing in the gas phase

greenhouse gas a gas that absorbs and radiates thermal energy in the atmosphere, effectively trapping warmth in the atmosphere

heat of fusion energy that causes the solid/liquid phase change without changing the temperature of the substance

heating energy transfer that increases the kinetic energy of a substance's particles

herbicide a plant poison

hydrocarbon a group of substances made of carbon and hydrogen

insoluble not capable of being dissolved

insulation material that can reduce energy transfers

ionic compound a compound in which atoms are connected to each other with a weak network of bonds (ionic bonds) rather than covalent bonds

kinetic energy energy of motion

lava molten rock flowing on Earth's surface

limiting factor the reactant that is present in the least amount in a reaction, which determines when the reaction stops

lipid a group of organic substances that includes oils, fats, and waxes

liquid a phase of matter that has definite volume but no definite shape. Loosely bonded particles in liquid can flow over and around one another.

mantle the large rocky part of planet Earth, located between the core and the crust

mass a measure of the quantity of matter

matter anything that has mass and takes up space

melt to change phase from solid to liquid

melting point the temperature at which a solid changes to a liquid; different for each substance

metal a group of elements that stretch, bend, and conduct heat and electricity

mixture two or more substances together that do not form a new substance

molecule a particle made of two or more atoms that are held together with strong (covalent) bonds

nitrogen a colorless, odorless, gaseous element that makes up about 78 percent of Earth's atmosphere

noble gas a gaseous element that does not react with other elements

nucleus the center of an atom, composed of protons and neutrons

octane an eight-carbon hydrocarbon molecule that is one of the main ingredients in gasoline

organic compound a substance produced by an organism

oxidizer a substance that provides an oxygen source for a combustion reaction

particle the smallest piece of a substance that is still that substance

periodic table of the elements an organization of the elements based on chemical properties

petroleum a natural resource made of mostly carbon and hydrogen; also known as crude oil

phase the physical condition of a sample of matter based on the kinetic energy of its particles. Common phases include solid, liquid, and gas. Also called state.

phase of matter the current state of a substance: solid, liquid, or gas

physical property a characteristic of a substance that can be observed without changing it chemically, such as size, shape, density, and phase

potash an impure form of potassium carbonate

precipitate an insoluble solid product of a reaction

predict to make an accurate estimation of a future event based on knowledge or pattern

product a substance produced in a chemical reaction

protein very large organic molecules that contain nitrogen

proton a subatomic particle that has a positive charge

pyrotechnics a field of science that studies explosive exothermic chemical reactions like those used to create fireworks

radiation a form of energy that travels through space

radioactivity radiation given off by the elements

ratio a mathematical relationship between two numbers

reactant a starting substance in a chemical reaction

room temperature the average kinetic energy of the particles in the air and other objects in a typical comfortable room

scanning tunneling microscope an instrument that can create images of arrays of atoms

solid a phase of matter that has definite volume and definite shape. The particles of a solid are tightly bonded and cannot move around.

soluble capable of being dissolved

solute a substance that dissolves in a solvent to form a solution

solution a mixture formed when one substance dissolves in another

solvent a substance in which a solute dissolves to form a solution

sublimation to change phase from solid to gas

substance a type of matter defined by a unique particle

symbol a representation of an element using specific letters

synthetic human-made

temperature a measure of the average kinetic energy of the particles in a substance

thermal energy radiant energy that heats

thermometer an instrument used to measure the average kinetic energy of particles in a substance

transparent matter through which light can pass and an image can be seen clearly

vacuum space containing no particles of air or anything else

vibrate to move rapidly back and forth

volume a defined quantity of space

water vapor the gas phase of water

well-ordered array a repeating pattern

Index

A
alchemy, 126, 186
atmosphere, 20, 70, 90, 96, 143, 186
atom, 110, 118–129, 155–157, 186
atomic number, 156, 186

B
Becquerel, Antoine-Henri, 12
biomagnification, 83, 186
bond, 29–30, 66–67, 93, 99, 186
burning, 124, 134–139, 186

C
calorie, 106–107, 186
carbohydrate, 22, 186
carbon dioxide, 20, 84–85, 119–120, 142–143, 146–147, 186
chemical equation, 120–129, 186
chemical formula, 26, 90, 120, 122, 146–147, 186
chemical name, 139, 186
chemical property, 5–8, 186
chemical reaction, 114, 119–129, 131–132, 135, 143, 149, 153, 186
climate change, 84, 186
combustion, 122–124, 131–139, 186
common name, 90, 186
compound, 114, 120, 153, 186
compressed, 31–32, 186
concentrated, 79, 186
concentration, 74–88, 186
condensation, 97–100, 186
conduction, 49–50, 56–59, 63, 186
conservation of energy, 48, 186
conservation of mass, 138, 186
conserved, 48, 107, 187
constraint, 58, 61, 187
contraction, 39, 40–45, 187
cooling, 39, 42–55, 108–109, 187
core, 17, 54, 187
criterion, 58, 61, 187
crude oil, 113, 187
crust, 17, 187
crystal, 94, 187
Curie, Marie and Pierre, 12
cyclotron, 157, 187

D
Davy, Humphry, 11
density, 34, 187
deposit, 100, 187
dilute, 78–79, 86, 187
dissolve, 19, 64–68, 70–73, 121, 187
dry ice, 100, 187

E
electron, 155, 187
element, 3, 25, 69–70, 81–82, 155–159, 187
Elion, Gertrude B., 152–154
energy transfer, 48, 60, 92–97, 102–109, 187
engineer, 41, 45, 57, 61, 63, 114, 187
engineering problem, 57–58, 187
equilibrium, 52–54, 103–104, 187
evaporation, 73, 75–76, 80, 96, 97, 99–100, 187
exothermic reaction, 131, 187
expansion, 38–39, 40–45, 51, 187
extinct, 87, 187

F
force, 29–31, 187
freeze, 89, 92, 94–95, 98, 100, 187
freezing point, 98, 99, 187
fundamental, 5, 187

G
gas, 11, 28–34, 37–40, 43, 62, 70, 74, 89–100, 136–139, 144–146, 187
gaseous, 29, 187
Ghiorso, Albert, 157
greenhouse gas, 84, 147, 187

H
Harris, James A., 159
Hayes, Tyrone B., 87
heat of fusion, 100–109, 187
heating, 35–39, 40–43, 46–55, 65, 75–76, 79, 84, 92, 97, 188
herbicide, 87–88, 188
Hitchings, George, 152
hydrocarbon, 144–145, 188

I
insoluble, 73, 121, 188
insulation, 57–63, 188
ionic compound, 188

K
kinetic energy, 35–43, 47–56, 93–97, 108–109, 188

L
lava, 89, 92, 94–95, 188
Lavoisier, Antoine-Laurent, 134–140
Lawrence, Ernest O., 157, 160
limiting factor, 133, 188
lipid, 22, 188
liquid, 14, 28–32, 37–40, 42, 58–59, 66, 79, 89–100, 105, 188

M
mantle, 17, 188
mass, 5, 67, 69, 96, 137, 138, 188
matter, 5, 20, 28–34, 54–55, 89, 121, 127, 138, 188
melt, 65, 78–79, 92–94, 97, 100, 103–105, 108, 188
melting point, 98–99, 188
Mendeleyev, Dmitry Ivanovich, 6–9
metal, 10, 13, 99, 188
mixture, 33, 68, 70–72, 188
molecule, 120, 141–143, 152–153, 188

N
Nelson, Donna, 149–151
nitrogen, 16, 188
noble gas, 10, 188
nucleus, 155, 157, 188

O
octane, 146–147, 188
organic compound, 141–147, 188
oxidizer, 132–133, 188

P
particle, 15, 24–27, 29–39, 47–55, 58–59, 66–67, 72–81, 90–104, 120–121, 125–129, 133, 157, 188
periodic table of the elements, 6–10, 82, 156, 188
petroleum, 112–117, 144, 188
phase, 20, 28–32, 34, 92–100, 188
phase of matter, 89–100, 188
physical property, 5, 188
potash, 11, 189
precipitate, 122, 126, 143, 189
predict, 8, 189
Priestley, Joseph, 136, 138
product, 112–116, 119–122, 150–151, 189
protein, 22, 189
proton, 157, 189
pyrotechnics, 132, 189

R
radiation, 12, 189
radioactivity, 12, 189
ratio, 77, 189
reactant, 119–123, 130, 142, 189
room temperature, 35, 54, 56, 59, 108, 189
Rutherford, Ernest, 155

S
scanning tunneling microscope, 156, 189
Seaborg, Glenn, 157
solid, 28–32, 37–40, 64–67, 71, 79, 89–100, 105, 189
soluble, 72, 189
solute, 68–80, 189
solution, 11, 68–79, 86, 125–127, 189
solvent, 68–80, 189
sublimation, 100, 189
substance, 4, 12–14, 25–26, 72–73, 81–82, 89–95, 98–100, 110–114, 119, 140, 149–150, 189
symbol, 6, 189
synthetic, 110, 115–117, 189

T
temperature, 41, 42, 46–47, 55, 189
thermal energy, 84, 131, 146, 189
thermometer, 14, 50–51, 55, 189
transparent, 68, 189

V
vacuum, 58–59, 62, 189
vibrating, 38, 50, 189
volume, 28, 32, 69–70, 189

W
water vapor, 33, 90–91, 96–97, 189
well-ordered array, 155, 189